9/18/90
David J. O'Brien

NEHA AKCAKAYA

Stacie
Connero
9/18/90

Nikos J. Ninos

PHIL WESTON

Annie S. Van Sickle

Cathy M. Shaw

ADVANCES IN CERAMICS • VOLUME 28

FAILURE ANALYSIS OF BRITTLE MATERIALS

ADVANCES IN CERAMICS • VOLUME 28

FAILURE ANALYSIS OF BRITTLE MATERIALS

V. D. Fréchette

The American Ceramic Society, Inc.
Westerville, Ohio

LIBRARY OF CONGRESS
Library of Congress Cataloging-in-Publication Data

Frechette, Van Derck, 1916–
    Failure analysis of brittle materials / V.D. Frechette.
       p.   cm. — (Advances in ceramics ; v. 28)
    Includes bibliographical references (p.
    Includes index.
    ISBN 0-944904-30-0
    1. Brittleness.   2. Fracture mechanics.   3. Glass—Fracture. I. Title.
  II. Series.
  TA418.16.F74   1990
  620.1 '126—dc20   90-1098
                                            CIP

ISBN 0-944904-30-0

Coden: ADCEDE

Printed in the United States of America.

1   2   3   4—93   92   91   90

Contents

Preface

This book is written to help those who are engaged in materials R&D, in manufacturing or in applications engineering, in geology, or in the forensic sciences and can profit from the ability to read cracks and fractures. Information about materials, quality of fabrication, and the service conditions leading to failure is there for anyone who is patient and observant enough to see.

The fracture itself is the best teacher, and indeed the study of cracks formed under controlled conditions is essential to acquiring proficiency in reading the signposts of failure. Nevertheless some fundamentals which should facilitate the learning process will be offered here.

Acknowledgement is gratefully offered to the many colleagues and former students with whom the author has had the privilege of working and of questioning the mysteries of crack behavior. Stimulating discussions are remembered with Carl Cline and William Snowden of the Lawrence Livermore National Laboratory, Professor Kulander of Wright State University, and with John Lonergan and his associates at Corning, Inc. Professor James R. Varner was kind enough to read the manuscript, and to make valuable suggestions.

Alfred, New York
June, 1990

Introduction

The science of fractography has arisen out of the necessity to learn why articles have failed during manufacture or in service, to see that laboratory test specimens have been stressed as planned, to identify unanticipated stresses which have arisen in engineering applications, and to reconstruct what testimony the fragments of glass, ceramics, plastics, or metals can contribute in cases at issue in law. It is the study of fragments and their interpretation in terms of material properties and conditions leading to failure. It is primarily concerned with the topography of the fracture-generated surface, including such zones near that surface as may have been associated with the event.

Originally, fractography embraced simply those observations which corresponded to everyday experience. A broken axle whose fresh fracture surface showed rusting over a portion of its area had obviously been cracked part of the way through, prior to its eventual failure. A window pane exhibiting a cone-shaped cavity surrounded by a few short radial cracks had evidently been struck by a small, hard, high-speed projectile. (Parents of children with BB guns will recognize the phenomenon!) Indeed, no one can doubt that Stone-Age man could readily distinguish between a flint tool knapped by pressure-flaking and one shaped by percussion.

The systematic description of fracture-generated surfaces originated with de Freminville in 1918. Interpretation of markings on glass crack surfaces owes much to Preston, to Wallner, to Smekal, and to Poncelet. Gilman showed how dislocation movement in a ceramic can initiate cracking and affect its development. Many others have published valuable results pertaining to the scientific analysis of failures in brittle materials; some of them will be cited in the appropriate sections. However, little is known about the many who have developed practical fractography in the course of engineering studies, corporate product development and control, and forensic investigation. No attempt will be made here to search for them, but it is certain that fractography owes much to studies which were motivated by an immediate need to know.

To a casual glance through the contents of this volume, it may appear as though it was written primarily for glass specialists. It was not, although it is hoped that they will find it useful. Glass is used primarily to typify all those brittle materials whose fracture is uncomplicated by such directional cracking behavior as cleavage (in single crystals) and parting (in rocks) or by a granular nature (such as found in concrete and ceramics).

Glass is the ideal material for study as the first stage in understanding failure analysis of all brittle materials. The lessons learned from glass can then be extended to include the effects of cleavage and granularity and so be applied equally to other brittle materials.

It is convenient to refer to stress systems in terms of everyday wares such as window glass, bottles, tableware, and the like because they are familiar and because their inexpensiveness makes them easily available for experiment. The engineer will not fail to recognize the principles applying in a pop bottle failure as identical with those operating in other pressure vessels, and to identify the stresses in a center-heated window pane with those in a computer chip or in a refractory slab. The geologist will have no trouble in identifying in rocks the very marks which are the principal signposts of fracture in glass.

FAILURE ANALYSIS OF BRITTLE MATERIALS

V. D. Frechette
Emeritus Professor of Ceramic Science
New York State College of Ceramics
Alfred University, Alfred, New York

1. The Initiation and Development of Brittle Failure

1.1 Brittleness

In everyday language, the term *brittle* is used as a synonym for fragile or lacking tensile strength. This association has some merit, but it is hardly justified in the case of the many modern brittle ceramics and glasses which are scarcely fragile and whose tensile strength is by no means trivial, with values as high as a million pounds per square inch.

A brittle failure is properly distinguished from a ductile failure or a viscous failure in that its fragments can be fitted together exactly, the reassembled specimen having precisely the same shape as before.

The term *brittle* is sometimes used by metallurgists in a relative sense to denote metals which show very little ductility. Such metals show the fracture characteristics that are to be described here. But none of the known metals is completely brittle and, to the extent that they exhibit ductile behavior, that is, continuous permanent deformation under shear stress, their fracture characteristics depart from those of truly brittle fracture.

From the mechanical standpoint, brittle materials can be defined as those which adhere strictly to Hooke's law behavior. They deform elastically, that is, reversibly, under applied load until the onset of cracking. Ductile materials, by contrast, show permanent deformation under load if strained beyond a certain point, called their *elastic limit*.

A few ceramic materials show slight ductility at room temperature. Most ceramics are ductile under very high confining pressure and also at high temperatures.

1.2 The Griffith Criterion

It is the very inability of the brittle material to deform inelastically that may lead to its failure under stress. Griffith,[1] following stress calculations by Inglis,[2] equated the energy required to create the two new surfaces formed in cracking with the energy expended in opening the crack, assuming that no other work was involved. The Griffith equation expresses the "notch sensitivity" of brittle materials:

$$\sigma_F = 2 \sqrt{[E \cdot \gamma / \pi \cdot c]}$$

where σ_F is the failure stress (strength), γ the surface energy of the material, E Young's modulus of elasticity, and c the depth of the starter flaw. In simple terms: a flaw in the surface of a brittle material reduces its strength dramatically.

From experiments with cracks of known depth, Griffith concluded that glass always contains inherent defects. He calculated the size of those in his own glass to have been in the order of 5 micrometers, assuming that they had the same shape and orientation as those which he deliberately introduced in his experimental series.

Unhappily, this has led to use of the term "Griffith cracks," by which some writers mean the undetected flaws in nearly pristine glass, while others mean those flaws large enough to grow catastrophically under a certain loading. Properly speaking, the Griffith crack size is the dimension of a hypothetical flaw of defined shape and orientation which would cause failure of a specimen under the same stress as actually measured in a particular specimen at failure. It may bear little resemblance to the flaw actually responsible for failure. (It is exactly analogous to the "Stokes diameter" of a particle settling in a viscous fluid; the Stokes diameter is the diameter of a sphere which would settle, other things being equal, at the same rate as the specimen particle is observed to settle.)

Fig. 1–1. Fracture markings consisting of twist hackle and arrest lines on a sandstone cliff in New Mexico. Cracking direction from bottom to top.

1.3   Life History of a Brittle Failure

In most of the instances in which we encounter failure of a brittle material—the breaking of a dish dropped on the floor, an over-speeded grinding wheel, or a bursting pop bottle, for example—we get the impression of an event so rapid as to be thought of as instantaneous. Indeed words such as "smashed," "disintegrated," "pulverized," "exploded," "splintered," often used to characterize such failures, are particularly used to describe sudden, instantaneous catastrophe.

There are other instances where we are able to see that a material failure develops from a particular site in stages, for example, when a crack in a windowpane is observed to extend stepwise farther and farther across the pane over a period of weeks or years.

The fact is that *in every case fracture begins at a particular site and grows from there.* Sometimes breakup may begin simultaneously from several sites and grow from each of them. The cracks may develop very slowly, at speeds that have been measured to be as low as a steady microinch per day, or may reach velocities of over a mile per second.

The crack may proliferate by forking, also called bifurcating or branching, and it may do so repeatedly, generating a family of many cracks sometimes resembling the branches of a tree.

A crack may decelerate as it runs and may come to a stop. Later it may restart and run to a new point of arrest, or it may continue to completion, separating the object into pieces. It may become complete by venting (running out at an edge), or by intersecting an earlier crack, or by looping around to intersect itself.

The crack may be assisted in its progress by the presence of certain substances in the environment, most particularly moisture, whether from the humidity of the atmosphere or from liquid water.

All of these phenomena are to be discussed in more detail. But here the important thing to recognize is that *a material's failure is never an instantaneous event but is always one which has a beginning, a more or less complicated history of development, and an end.*

The crack runs straight, it curves, or it meanders. It advances along a straight front to generate a flat surface, or it cups toward the middle or curls at one edge. The crack front may be perpendicular to the specimen surface or inclined to it. The crack surface may develop smoothly; at times it may be rough. It may show a variety of markings, some riblike or ripplelike; some like the wings of a bird; some like the profile of a mountain range; some like spears, barbs, or sprays. All of these markings may occur solely as the result of the environment, the stresses which drive the crack, and the shape characteristics of the specimen. Microstructural features may further complicate the course of crack development.

Conversely, the crack surface contains detailed information about the forces that were at work during its growth. They can be interpreted later to reconstruct details of the stages in which it developed to gain valuable clues as to the nature of a particular material's failure.

1.4   The Matter of Scale

One of the most dramatic features of cracking phenomena has to do with the matter of scale. The predominant features of crack topography are to be found over an enormous

Fig. 1–2. Transmission electron micrograph showing crack markings on grains of a BeO ceramic, including twist hackle (sheaves of dark lines) and Wallner lines (arc-shaped). Cracking direction from lower left to upper right. × 20,000.

range of sizes, from the smallest scale observed under the electron microscope to the greatest scale seen on cliffs a mile high. Figure 1–1 shows twist hackle and arrest lines on a cliff photographed at a magnification of × 1/1000, that should be compared with Fig. 1–2, a single grain of a cracked ceramic photographed at × 20,000.

2. The Fundamental Markings on Crack Surfaces

2.1 The Origin Flaw

The strengths of glasses, typical of brittle solids, are in simple theory so high as to make design engineers dream. The real strengths—the engineering strengths—are very much lower. To begin with, such "theoretical" strength is based on the force necessary to break chemical bonds, assuming that the applied stress is borne uniformly over all such bonds. It is not. If an atom is missing (there are missing atoms in virtually all solids) or if there is a misfit atom (if there is no missing atom there will be extra atoms), the applied stress is not supported equally everywhere but is concentrated about such defect sites. Failure begins in such zones and, as they fail, bordering regions must bear the still more enhanced stresses and they too fail. By this means, theoretical strengths are reduced by about an order of magnitude. Thus vitreous silica, with a theoretical strength of ~~much~~ about ten million pounds per square inch, has never shown a measured strength of more than a million pounds per square inch, even under the most carefully controlled laboratory conditions. "Theoretical" strength is not only unachievable in practice but, when theory is properly applied, it is seen to be theoretically unattainable also.

Practical strengths are still more limited. Exposure to the atmosphere alone can degrade strength. The lightest contact with the softest of materials reduces it. Ordinary handling further degrades strength and rough handling reduces it still more.

Such processes of degradation take place at exposed surfaces, and it is for this reason that it is the surfaces of homogeneous brittle materials that are so vulnerable to crack initiation. A discontinuity in the surface in the form of a notch, a groove, a microcrack, or simply an atomically weak site acts as a stress concentrator, or "stress raiser," that is, it acts to concentrate applied stresses at its boundary and these concentrated stresses can initiate failure.

The facture origin is that flaw (discontinuity) from which cracking begins. Its importance is paramount in failure analysis. If the crack had not begun, obviously failure would not have occurred.

The discontinuity, whether brought about by chemical, thermal, or mechanical agency, has its own history of development and will bear the record of that development on its own surface, whether or not this is always readable by contemporary means. In the case of discontinuities large enough to reduce strength below the engineering range (for glass this is about 5000 pounds per square inch, depending on the application), such crack origins can usually be identified as to source, whether from impact, abrasion, or process defect. Indeed they are analyzed by the same principles used in inferring the history of crack development. These principles will be developed in the following discussion.

Fig. 2–1. Arrest line on the crack surface of flat glass. The profusion of tertiary Wallner lines preceding the arrest (left) is in marked contrast to the smoothness of the surface generated on resumption of cracking. Cracking direction left to right. Differential interference contrast; × 100.

2.2   Signposts of Failure Analysis

As it runs, the crack develops a new surface which is perpendicular to the axis of principal tension at its tip. This fact is central to understanding the topography of the crack surface and to discussing the markings which may appear on it.

It is also necessary to state, broadly at this point, that *the higher the stress at the crack tip, the faster the crack will run,* up to some effective terminal velocity. The details of this relationship will be discussed later.

At a particular instant, and in a particular part of a running crack front, the axis of principal tension may be pitched forward, or it may be rolled laterally, owing to the action of some transitory or persistent influence. The consequence of such a shift is to modify the plane of cracking. A close analogy is the cutting of a groove in a phonograph record. In the absence of an external agency, the groove is cut smooth and featureless. But if a sound is made, deforming a diaphragm and perturbing the force on the cutting needle, the groove being cut will contain surface irregularities which record the sound in accordance with its tone, pitch, and intensity. These irregularities can be seen under the microscope. They can also be scanned to reproduce the precise sonic spectrum of the sounds that caused them.

A considerable variety of markings may be seen on crack surfaces of an essentially

Fig. 2–2. Contour lines of a crack surface showing the wavelike cross section of the Wallner lines. Kohaut-Michelson interferometer; × 100.

homogeneous brittle solid such as glass. All of them observed so far can be interpreted as variations of just a few fundamental types and subtypes. One or two must be added in the case of single crystals and polycrystalline and polyphase materials; these materials will be described in a later section.

One family of markings, called by the general term "rib marks" to describe their shape, are almost always concave in the direction from which the crack was coming. They include arrest lines, three kinds of Wallner lines, and certain scarps, each of which has particular information to convey in unraveling the history of the fracture.

Four kinds of hackle, whose component lines run parallel to the local direction of crack spreading, add information about crack velocity, shifting stress fields, and structural inhomogeneities.

Scarps, or escarpments, appear as subtle lines of various shapes which separate that portion of a crack surface which has been formed in the presence of a liquid from an adjoining portion formed in the absence of such liquid. At the very least, they are useful in identifying the presence of a liquid in a particular cracking event, and they also may indicate crack velocities and stress gradients.

The nomenclature of fracture markings has been confused, both in the scientific literature and in industrial usage. The present system has been assembled[1] primarily to group markings logically according to their nature and significance, when possible to

Fig. 2–3. Primary Wallner lines formed by encounter of the crack front with a complex of surface flaws (bottom left). Crack ran from left to right. × 250.

reflect credit on those who have contributed to their understanding, and in some cases merely to provide easily remembered names. Each term will be introduced with a formal definition which will be useful in testing doubtful cases, followed by illustrations to aid in its sight recognition, and by some discussion of its importance.

2.3   Rib Marks

RIB MARK

A riblike curved line on the crack surface, usually concave in the direction from which the crack is running.

This is a generic term, useful only for referring to a rib mark whose classification is unknown. Rib marks may be arrest lines, Wallner lines, or certain scarps, each of which has special significance and so should be called by its proper name as soon as it is identified.

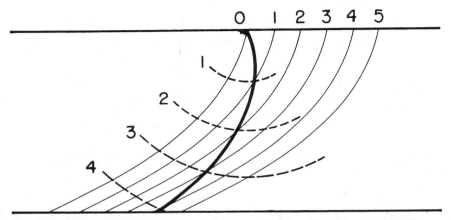

Fig. 2–4. Stages in the generation of a primary Wallner line. Encounter with a singularity at the free surface (top) initiates an elastic pulse. Light arcs in the sketch show positions of the crack front at successive times; dashed lines show corresponding positions of the (faster) elastic pulse. The locus of intersection forms the Wallner line, marked in bold face.

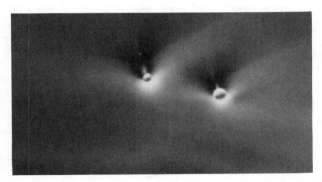

Fig. 2–5. "Gull wings," i.e., primary Wallner lines, each pair formed by encounter of the running crack with an inclusion. ×500.

2.4   Arrest Lines

ARREST LINE (also called dwell mark)
A sharp rib mark defining the crack front shape of an arrested crack prior to resumption of cracking under a more or less altered stress configuration.

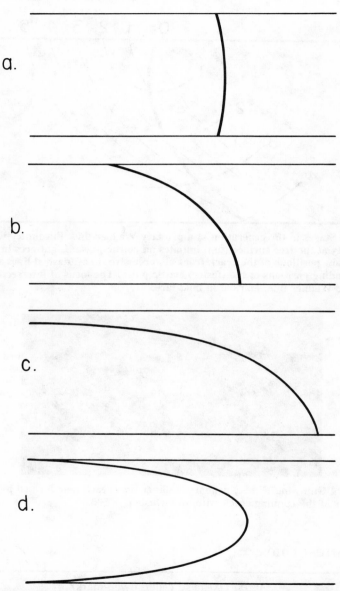

Fig. 2–6. Inference of stress distribution across a thin specimen from Wallner line shape: (*a*) **Tensile stress uniform across the thickness;** (*b*) **tensile stress highest at the lower surface;** (*c*) **tensile stress highest at the lower surface; upper surface under initial compressive stress;** (*d*) **tensile stress highest at the midplane.**

Fig. 2–7. Primary Wallner line formed by breakthrough of the crack to the second free surface (top) and extending forward, from upper right to lower left, to reach the first surface (bottom), at which the crack was leading. In doing so it has intersected a family of secondary Wallner lines, each initiated by a detail in the mist hackle at the lower part of the surface. × 100.

The term *arrest* should not be taken too literally. Indeed, the arrest line may not involve a halt in crack spread. The important feature is that it represents the site of a cracking discontinuity, of a more or less radical nature, in the dynamic stress system. This may indeed have involved an arrest, or halt, in the spread of cracking; it is then virtually inevitable that the axis of principal tension reinitiating cracking should differ from that which preceded arrest, and a trace of the crack front is left on the crack surface. But a sudden shift in the axis of principal tension can also result from an abrupt shift in the stress distribution. In this case the shape may be distorted from that of the crack front.

The time elapsing before the crack continues cannot be learned from the arrest line itself, although other markings in the vicinity may hint at the duration of the arrest. Figure 2–1 illustrates an arrest that may have been of considerable duration, since the crack-surface segments before and after arrest differ so markedly.

Recognition of an arrest line as such is not always easy. If just a forward tilt of the principal tension axis is involved, the arrest line will be characterized only by the abruptness of its contour; this distinguishes it from the primary Wallner line (see below) as does the absence of the perturbing detail necessary to generate a primary Wallner line. The arrest line differs from the cavitation scarp by the absence of precavitation hackle preceding it.

Fig. 2–8. Secondary Wallner lines generated by encounter of the crack front, moving left to right, with mist-hackle roughness (along the bottom, dark in the photo). × 100.

If a twist in the axis of principal tension occurs at the arrest line, twist hackle will trail downstream from it; its identity is then immediately recognizable.

2.5   Wallner Lines

WALLNER LINE (general)
A rib-shaped mark with a wavelike contour caused by a temporary excursion of the crack front out of plane in response to a tilt (a pitch, or fore-and-aft shift) in the axis of principal tension. (It may also result from passage of the crack front through a locally shifted stress field, as at an inclusion or discontinuity at a free surface.)

A Wallner line is the locus of intersection of a spreading elastic pulse with successive points along the running crack front. Because the Wallner line is the record of passage of an elastic pulse, its profile is determined by the shape of the pulse, that is, the history of its rise and decay in passing a point. Since the elastic pulse is typically wavelike, the resulting Wallner line profile is also wavelike (Fig. 2–2), and it is this which distinguishes it from an arrest line or other rib mark.

Three types of Wallner lines fit the definition. Following Wallner's classification,[2] the first two are designated primary and secondary. The third type was not discussed by Wallner, but it conforms to his explanation of the first two and it is included here under his name for the sake of consistency.

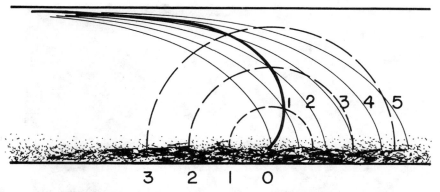

Fig. 2–9. Stages in the production of a secondary Wallner line, caused by mist hackle roughness at the lower edge. Numbered arcs in the sketch show positions of the crack front at successive times; dashed lines show corresponding positions of the (faster) elastic pulse generated at one of the roughness details. The Wallner line is the locus of their intersection.

From time to time, in discussing Wallner lines, it will be necessary to mention velocities. While crack velocity is to be discussed later, under the heading of Mist Hackle, it should be noted here that the effective terminal velocity of a crack is about 60% of the velocity of shear waves in the material. For soda-lime glass, that is equivalent to a terminal crack velocity of about a mile per second. This means that typical fracture events in specimens of modest size are completed in less than a millisecond.

Primary Wallner Line

A Wallner line formed by an elastic pulse generated by encounter of some portion of the crack front with a singularity in the specimen, such as a discontinuity at the free surface or within the specimen, or with any localized stress field or elastic discontinuity (Figs. 2–3, 2–4, 2–5).

Because of the obvious resemblance of the pair of primary Wallner lines in Fig. 2–5 to the wings of a soaring bird, they are commonly referred to as **gull wings**. Winglike primary Wallner lines may originate at any one of several points in passage by an inclusion and they might be given other names, depending on their location, but all are simply special cases of primary Wallner lines.

The primary Wallner line is useful in inferring the direction of cracking. It is also especially valuable for the information it gives about scratches, bruises, and other surface flaws seen at the edge of a crack in postmortem examination. If a Wallner line starts in such a flaw, that flaw can be concluded to have been present at the time of cracking (Fig. 2–3). If, on the other hand, no Wallner line was initiated at the intersected flaw, it was evidently introduced after the incident.

Primary Wallner lines are sometimes the only identifiable markings in cases where the crack runs unusually smoothly or where the granular nature of the material makes

Fig. 2–10. Tertiary Wallner lines, caused by interaction of vibrations, externally generated by impact, with the crack front, expanding from the center of impact at the lower left in the photo. × 100.

other markings difficult to see. Then a design feature in the specimen should be particularly examined to see if it has generated a Wallner line (sometimes it is a broad ridge) in the surface and thereby pointed the direction of cracking and the shape of the crack front (Fig. 2–6).

A primary Wallner line of special interest is formed at the breakthrough of a crack to the second free surface (Fig. 2–7). It should not be confused with an arrest line. Its particular value is its indication of the distribution of tensile stress about the origin at the moment of failure. Uniform tensile stress through the thickness is indicated by a primary Wallner line that is semicircular about the origin, whereas tensile stress that is decreasing away from the origin through the thickness develops an elliptical one.

SECONDARY WALLNER LINE

A Wallner line generated by an elastic pulse released by a discontinuity in the progress of the crack front, typically one of the rough details which arise as the crack approaches its effective terminal velocity (see Mist Hackle, below).

The typically hook-shaped secondary Wallner line is a striking demonstration of the fact, true of all classes of Wallner lines, that *the Wallner line does not indicate faithfully the shape of the crack front at its location.*

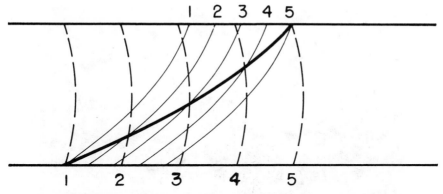

Fig. 2–11. Stages in the formation of a tertiary Wallner line caused by an externally generated elastic pulse. Numbered arcs in the sketch show positions of the crack front at successive times; dashed lines show corresponding positions of a single overtaking (faster) elastic pulse. The locus of their intersection is marked by the Wallner line.

The reason for this should now be explained. The fastest portion of the crack front is in the area where mist hackle appears, and therefore the crack front must be farthest advanced there. Yet the Wallner line is not most advanced in the mist hackle region of Fig. 2–8 but rather some distance removed from there, where logic and measurement indicate that the crack front lagged somewhat behind. The explanation is in Fig. 2–9, where it is clear that the crack front has advanced some distance since the Wallner line began before the spreading elastic pulse has had time to intercept it. Thus the locus of intersection of crack front and elastic pulse is not identical with the crack front, and must not be expected to be.

Since the secondary Wallner lines always accompany the formation of mist-hackle roughness that is associated with terminal velocity, one is apt to wonder if they themselves have any information to contribute. Their value is in those cases where the mist hackle is confined to a band so close to the free surface that it may have escaped notice. The presence of secondary Wallner lines, easily recognized by their fish-hook shape and by their typical abundance, call attention to its presence, while the shapes of the secondary Wallner lines indicate the steepness of the stress gradient through the thickness and possibly the presence of initial compressive stress at the trailing edge. (This is discussed in detail under the heading of Twist Hackle.)

TERTIARY WALLNER LINES
Wallner lines caused by elastic pulses generated from outside the crack front.

Mechanical shock or vibration imparted from an external agency are among those which may be responsible for the pulse. Mechanical shock, such as produced in impact with a hard surface, typically results in a ringing of the specimen which, though short-

Fig. 2–12. Wallner line "lambda mark" in thin sheet glass. Reflection of the sonic signal from the second surface gives a discontinuous extension of the Wallner line. × 250.

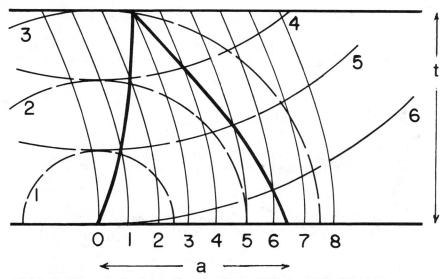

Fig. 2–13. Stages in the formation of a Wallner line lambda mark. Light arcs show successive positions of the crack front; dashed lines show corresponding positions of an elastic pulse, generated in an edge singularity, and its reflection in the second surface. The lambda mark is the locus of their intersection.

lived, persists throughout the primary breakup of the specimen, and so generates tertiary Wallner lines on most of the crack surfaces (Figs. 2–10, 2–11).

Vibration can also be set up by the sudden stress relief which occurs at the onset of cracking, as shown by the experience that the cracking event is usually audible as a snap. This vibration is brief, however, and the train of comparatively feeble elastic pulses is soon attenuated. Moreover, it quickly overtakes the crack front and passes beyond it, so that the tertiary Wallner lines generated in this way are characteristically feeble and extend only a comparatively short distance from the fracture origin. This is why window panes cracked by thermal means yield crack surfaces which, except those close to the origin, are quite devoid of tertiary Wallner lines and, indeed, may be entirely featureless.

Since Wallner lines are formed by the action of an elastic pulse with the running crack front, it is not unexpected to find that an elastic pulse can continue to form a Wallner line after reflection from a free surface. Figure 2–12 shows an example of this "lambda mark" and Fig. 2–13 traces the steps in its development. In the time that the lower tip of the crack front has traveled the distance a in the diagram, the elastic pulse has traveled twice across the thickness, t, of the sheet at the reflecting angle. Thus the crack velocity V_c is given by

$$V_c = V_s \cdot a/\sqrt{(a/2)^2 + t^2}$$

where V_s is the velocity of sound for transverse waves in the medium.

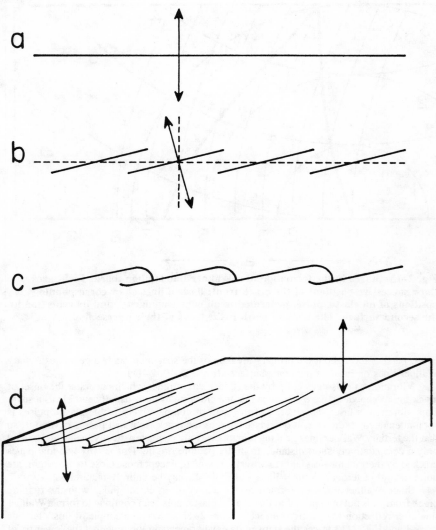

Fig. 2–14. Stages in the formation of twist hackle: (*a*) Cross-sectional view of the crack surface, forming perpendicular to the axis of principal tension and running away from the observer; (*b*) a lateral twist in the tension axis causes the crack surface to split into unconnected segments, each perpendicular to the new axis of tension, the average crack surface coinciding with the original; (*c*) breakthrough between the segments completes the separation of the specimen into two fragments, forming shallow "risers" in the stairway in which the initial cracks segments are the "treads"; and (*d*) isometric view of the resulting surface.

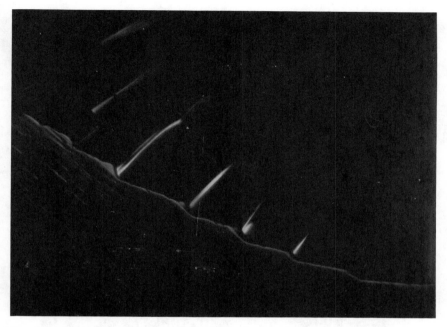

Fig. 2–15. Specimen sectioned across a spray of twist hackle revealing a staircaselike crack surface whose broad treads were generated by the advancing independent crack elements, and whose shallow risers were formed by breakthrough between them. × 120.

2.6   Hackle

HACKLE
A line on the crack surface running in the local direction of cracking, separating parallel but noncoplanar portions of the crack surface.

TWIST HACKLE
Hackle separating portions of the crack surface, each of which has rotated from the original crack plane in response to a lateral rotation or twist in the axis of principal tension.

Figure 2–14 shows the way in which twist hackle is formed. When a crack surface, initially developing perpendicular to the original axis of principal tension, encounters a sidewise twist in the tension axis, it splits into segments, each of them perpendicular to the new tension axis. At first these segments are not interconnected and the specimen is

Fig. 2–16. Breakthrough between the crack surface segments of hackle can occur from above (left), from below (center), or from both directions, separating a sliver (right).

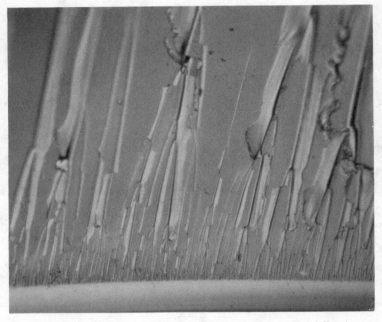

Fig. 2–17. Twist hackle generated upon resumption of cracking following an arrest. The crack is traveling from bottom to top. × 100.

not separated. Separation, demanded by the increasing strain on the system, is accomplished by lateral breakthrough from one segment to another. Figure 2–15 is an over-the-edge picture of a flat crack broken into a set of steps by this process. Its resemblance to a broad staircase is striking, the advance crack-surface segments constituting the treads and the breakthroughs the risers.

Breakthrough between steps may itself generate twist hackle on the "risers," a phenomenon which can be termed secondary twist hackle. The secondary hackle, in turn, can form twist hackle on its own breakthroughs, and this is tertiary twist hackle. Secondary and tertiary twist hackle has, in fact, been observed on large specimens, and further stages would not be astonishing.

Figure 2–16 shows how breakthrough may take place toward one direction or the

Fig. 2–18. Twist hackle forming at the trailing edge of a crack front, in a region which was under compression when the leading edge of the crack passed through, traveling from left to right. The hackle itself traveled from bottom to top. × 100.

other. Sometimes breakthrough occurs in both directions, and this separates slivers, often long after the main fracture event. This can be observed by listening to the debris for the tinkle that accompanies the slivering.

A switch from one lateral direction of breakthrough to the other is common, generating barblike features in the hackle pattern (Figs. 2–17, 2–18). The barbs are not to be mistaken for crack direction indicators, however; they may point either in the direction of cracking or toward its source.

Figure 2–19 shows the shape of the crack front in the advancing crack segments. As can be inferred by the tertiary Wallner lines in the photo, each segment leads in the center and lags at its edges. It is tempting to think that this means that the hackle lines (the "risers") are somehow retarding the crack front, but the effect is actually the natural consequence of partial overlap of the crack segments. This has the effect of partitioning the stress between the segments. Thus the stress is reduced in magnitude at each of them near the overlap, and progress is retarded accordingly.

While twist hackle faithfully blazes the local trail of fracture, it not always easy to interpret as to sense, that is, the direction in which cracking proceeded. Fortunately, twist hackle frequently starts out as a closely spaced array whose members run together, the shallow steps between the original individuals growing into higher steps by combination. From this tendency the term *river pattern* has arisen; it is as though small tributary rills ran together to form streams and the streams joined to form rivers. Such a configuration shows the sense of crack propagation unambiguously.

Just as river systems tend to diverge once they have become mature, the major hackle

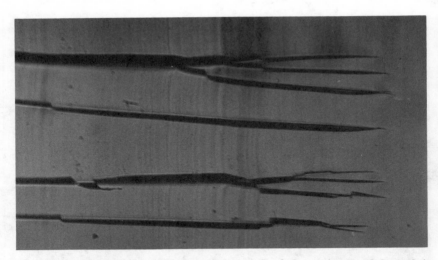

Fig. 2–19. Barbs on hackle surfaces are the result of alternating breakthrough in one direction and then the other. The tertiary Wallner lines in the photo show that the initial independent crack segments were retarded at their edges, where the overlapping segments provided compliance and thereby reduced the effective tensile stress at the crack fronts. × 500.

markings also tend to diverge, and they do so unless prevented by geometric factors or restrictive stress configurations. Also, the same sort of agency may sometimes cause major hackle markings to split, in very much the same way that twist hackle was generated in the first place. If the image of river patterns is preserved, they could be thought to constitute deltas. Tributaries can be distinguished from deltas by the angles at which junction occurs. A tributary tends to run more and more perpendicular to the river into which it joins, whereas the angle subtended by delta branches increases, initially, with distance from the junction.

SHEAR HACKLE
A particular form of twist hackle occurring in the lazy loops generated in the later stages of fracture of a hollow specimen.

Far from the fracture origin, it is common to see the crack turn through 90 degrees and develop a cupped surface, inclined about 45 degrees to the free surface, from whose center line a spray of twist hackle emerges (Fig. 2–20). It has little significance in reconstructing a failure event in industry, but it is useful in problems of rock fractures.

WAKE HACKLE
Hackle mark extending from a singularity at the crack front in the direction of cracking.

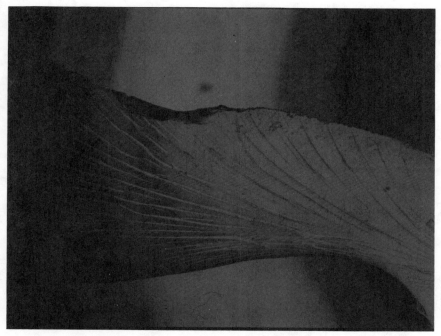

Fig. 2–20. "Shear hackle" in the latter stages of cracking from the rim (right) of a thick-walled glass jar. The addition of shear forces to the tensile stress inclines the crack surface about 45° from the free surface and forms a fan of twist hackle about the center line. × 30.

When an inclusion whose mechanical properties differ in any way from those of the host is encountered, the crack front is likely to divide at the inclusion and sweep past it on both sides. As the two fronts approach one another behind the inclusion they generally do not meet, but proceed onward parallel to one another but slightly overlapped. Eventually breakthrough between them occurs in the same way as described for twist hackle. The result is a single hackle step, trailing from the inclusion in the direction of crack spread (Fig. 2–21). Its resemblance to the wake seen downstream from a post in flowing water suggests its name.

Wake hackle may persist for a long distance on the crack surface, or it may fade out rapidly, that is, its step height may diminish rapidly, as the two crack fronts approach one another and finally merge into one.

As with twist hackle, breakthrough can occur to either side and it can switch from one side to the other, generating a barblike detail. Again, breakthrough to both sides will generate a splinter.

A void, pore, or bubble is a special case of inclusion and may also cause wake hackle (Fig. 2–5). It may seem curious that a void can interfere with crack spread, inasmuch as it already represents a separation of the material. But the relatively rounded shapes

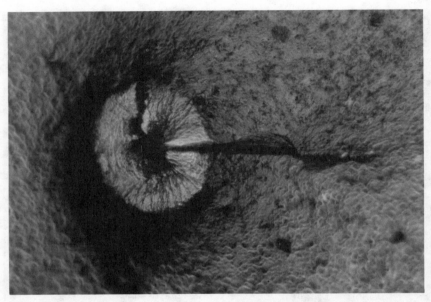

Fig. 2–21. Wake hackle extending from an inclusion in BeB$_6$ to the right, i.e., in the direction of cracking. ×250.

common to most voids prevent them from concentrating stress sufficiently to advance cracking as the sharp crack tip is able to do. Therefore, cracking downstream from the rounded void proceeds only when the sharp-tipped crack arrives there by passing on either side of the void.

Even sharp-edged inclusions may generate wake hackle, or at least a ridge on the crack surface indistinguishable from wake hackle, because the sharp edge is most unlikely to be aligned with the crack front. Additional effects at inclusions are discussed later.

MIST HACKLE

Markings on the surface of an accelerating crack close to its effective terminal velocity, observable first as a misty appearance and with increasing velocity revealing a fibrous texture, elongated in the direction of cracking, and coarsening up to the stage at which the crack bifurcates.

The crack velocity increases with increasing stress at the crack tip, forming, in the absence of perturbing influences, a smooth crack surface. Eventually, at higher velocities it begins to develop microscopic roughness, at first resembling a mist such as that observed when moisture condenses on a mirror (Fig. 2–22). At this point, the rate of increase in velocity with increasing crack-tip stress begins to diminish. As velocity increases further, a fibrous-appearing structure can be observed, elongated in the direction of cracking. At still higher velocity, hackle features appear within the misty surface, and

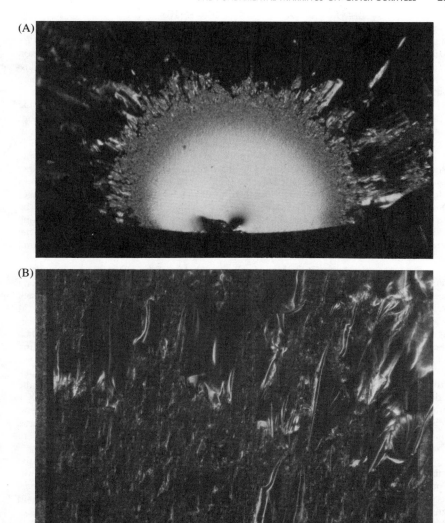

Fig. 2–22. (*A*) **Mist hackle in a semicircle about the origin of cracking of a glass rod broken in cross bending. Although the first visible roughness appears like a mist at low magnification (×40), at higher magnification, (*B*) (×500), it is seen to be textured.**

slivers may separate. The effects become increasingly coarse, the velocity becomes constant, and the crack may then undergo *velocity forking, velocity bifurcation,* or *velocity branching.* (These terms are used to distinguish the phenomenon from simple nucleation of a new crack from the edge of an earlier one.)

The surfaces of the two velocity forks lack mist hackle, but they may show wake hackle trailing from the rough surface features of the parent crack.

The crack velocity in the constant-speed regime preceding velocity forking is the effective *terminal velocity* under all but extraordinarily fast loading. This effective limit is imposed by the inability of the crack to maintain continuity along its front at speeds in excess of about half the speed of sound in the material.

On a microscale, even at very low speeds, there are local deviations from the rule that the plane of cracking is to be perpendicular to the axis of principal tension. In glasses, where long-range order is lacking, and bond strengths are consequently variable from place to place, the crack tip can be expected to deviate out of plane to separate weaker bonds in preference to stronger ones. But the extent of deviation permitted will be restricted by the energetic cost of generating additional crack surface. To express the same idea in terms of stresses, a local excursion from the general crack plane repartitions the stresses in the vicinity in such a way as to restore the deviated portion back to the general crack plane.

The same rule applies at the higher velocities, but it can be followed only as soon as the stress configuration is perceived at any point on the crack front. The interaction among locally perturbed stress fields is limited by the speed of sound in the material. At low crack velocities this is adequate to maintain energy-efficient continuity along the front. At higher velocities there is an appreciable delay between any local deviation and the response by its neighbors to accommodate it. Accommodation may therefore occur when the occasion for it has passed, and obsolete adjustments take place along the front, leading to undulations which are amplified reactions to local variations on the atomic scale.

At still higher velocities, undulation builds and continuity may be lost. The crack front becomes an array of tonguelike elements, each more or less perpendicular to the principal tension. Some of these tongues deviate so far from the average plane that they are aborted when their more nearly in-plane neighbors close in together beneath them; mismatch, as they meet, yields wake hackle. Figure 2–23 shows a number of these microcracks in profile at the free surface. Others are to be found in any section parallel to the free surface.

The formation of cracks aborting to each side of the crack plane, together with the mismatched joining of separated crack elements and similar departures from a smooth, continuous crack front, dilutes the concentration of tensile stress among them and serves as a brake on further acceleration. Additional stress on the system does not result in greater crack-tip stress but merely generates still more aberrations during crack advance. Aborting cracks run progressively deeper until finally one of them continues to run as a mature crack fork, and velocity bifurcation of the parent crack has then become realized.

The preceding argument implies that the effective terminal velocity of crack propagation is set by the velocity at which velocity forking occurs and this in turn is set by the speed of sound in the material and by structural characteristics which provide preferential paths for local crack extension.

It should be noted that the practical limitation on crack velocity has not always been explained in this way. A number of theories ascribe it to diverse physical effects, includ-

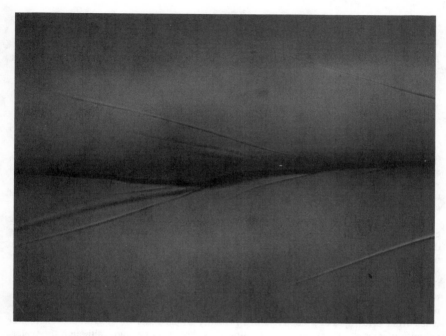

Fig. 2–23. Free surface of a glass specimen broken in cross bending and reaching terminal velocity. Microcracks are seen branching from the mist hackle region of the main crack and aborting. The details have been accentuated by a 10-s exposure to HF vapor. ×100.

ing the dispersion of Rayleigh wave velocities. All such theories imply that it is quite impossible for the crack to exceed some fixed value, variously calculated to be between 50 and 61% of the transverse speed of sound. In fact, however, values of over 95% were measured by Snowden,[3] when he loaded the specimen very suddenly. Under these circumstances no mist hackle appeared on the surface and forking did not always occur. His observations are consistent with the argument presented above. In his experiments there was neither time nor space to develop instabilities at the crack front. Crack velocity was high at the outset, because there is no inertia in a crack and therefore no need to go through an acceleration phase. At all points along the crack front extension was regulated by the principal tension alone, without deviation to seek weaker paths in the material and without perturbing effects from neighboring segments.

2.7   Scarps

SCARP

A line on the crack surface, which is the locus of intersection of a liquid-filled part of a running crack front with an unwetted part, or of a moist part with a dry part.

Scarp formation is understandable in terms of the effect of water, as liquid, as vapor, or in solution. At low levels of applied stress, water has a profound influence on crack spread. Indeed it can permit cracking which would not occur at all in dry air or in high vacuum. Figure 2–24 shows the relationship. Water is seen to lower the stress required to run a crack at a particular velocity in the slow-growth range, that is, up to about 0.005 m/s (2 in./s).

As crack velocity increases in the slow range, the aid of liquid water in promoting crack growth is offset more and more by the viscous resistance of the column of water dragged through the narrow crack opening by capillary forces at the running crack tip. The water column experiences two opposing forces, the one pulling it forward, the other resisting. Thus a negative pressure is developed within it. Only the cohesive strength of the liquid prevents separation, that is, cavitation, before the negative pressure developed within the liquid exceeds a value known as the cavitation threshold. When this value is reached in the running crack, cavitation occurs at the tip and the viscous drag ceases. The crack front dries up and velocity jumps instantly by an order of magnitude. Thereafter velocity increases with stress intensity almost as though in vacuum (although the water, still following slowly, exerts a small drag and so reduces the velocity slightly).

As in all nucleation and growth processes, cavitation nuclei can form spontaneously or with the aid of foreign surfaces, the more so as velocity (and negative pressure) increases. At first they are too small to grow and they promptly collapse. Now, the threshold cavitation pressure is strongly reduced by the presence of dissolved gas in the liquid,[4] so the early precavitation events are marked by air being depleted out of the solution. During the short lifetime of such an air bubble on the surface, it constitutes a dry spot at the crack tip. The crack cannot grow at such a site under the stresses effective there. The crack is therefore arrested at the small sector of the crack front affected. Neighboring wet portions on either side sweep around it and close together beyond it, isolating a small pocket of air and leaving it behind. As in the case of passage of a crack past any inclusion, a wake hackle step is formed.

As in the case of hackle in the wake of an inclusion, the wake hackle downstream from a subcritical cavitation nucleus may eventually smooth out and disappear. Increasing velocity enhances the probability of subthreshold cavitation, and the density of wake hackle on the crack surface increases until either all the air is exhausted from the liquid, a state which requires a lengthy period of slow acceleration, or until the stable cavitation threshold pressure is exceeded. Cavitation then occurs and a **Michalske scarp** or **cavitation scarp** forms, as first described and explained by Michalske.[5]

This occurs at a velocity of about 0.04 m/s for soda-lime-silica glass at room temperature. The cavitation scarp itself has no distinguishing feature, but it is recognizable by the sheaf of very fine wake hackle (**subcavitation hackle**) which almost always precedes

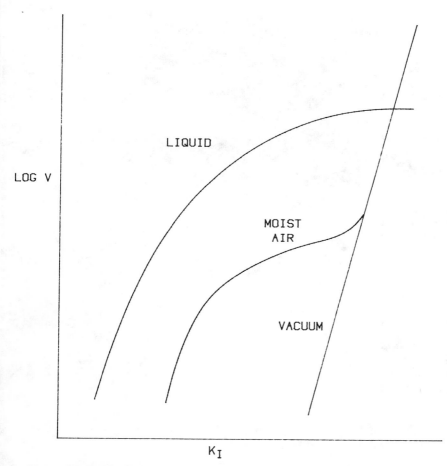

Fig. 2–24. Relationship between stress intensity, K_I, and velocity of a running crack. Moisture from the air lessens the stress necessary to run a crack at a particular velocity; liquid water lessens it still more. × 500.

it (Figs. 2–25, 2–26). The hackle ceases at the scarp, or it may disappear earlier, as in cases of very slow acceleration or when the water is poor in dissolved gases (as it is when recently boiled or when it is the product of melting ice). Subcavitation hackle does not occur at all in later cycles of periodic acceleration and deceleration when the air has been stripped out in previous cycles.

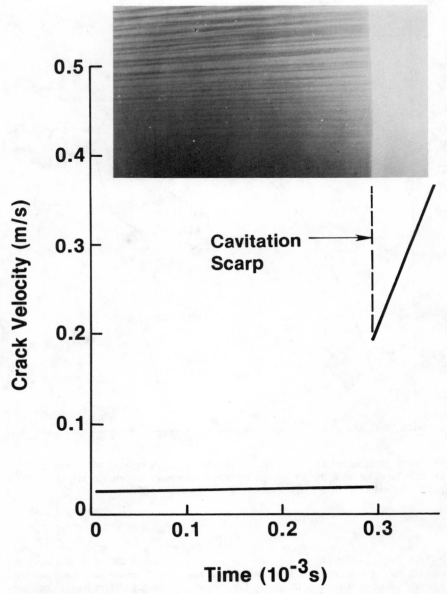

Fig. 2–25. Subcavitation hackle terminating in a cavitation scarp. Crack direction left to right. ×100.

Fig. 2–26. Cavitation scarp at which a line of twist hackle begins. Subcritical cavitation hackle is seen in the lower part of the photo. Crack direction bottom to top. ×500.

The cavitation scarp may take on various appearances, ranging from a nearly invisible line to one which is marked by twist hackle formation (Fig. 2–26). The reason for the scarp itself is not known with certainty, but its significance is clear: Its appearance, following a region of spontaneous fine wake hackle is always evidence of transition from a wet-running crack to the dry regime.

It could be imagined that an analogous scarp would be observed when a crack decelerates through the critical velocity. The conditions are not reversible, however. Because there is no way in which water can be ingested instantaneously along the crack front of a decelerating crack, deceleration generates scarps of other kinds. If water is available at the free surfaces, its influence is first felt there. Water is drawn in by capillarity at the edges and there it retards crack growth by its viscous drag. The wetted portion is therefore at first retarded with respect to the dry, as indicated by timing marks imposed on experimental systems. As the crack slows further, a velocity is reached—the *equivelocity point*—at which water, intruding ever deeper along the crack front, neither facilitates nor retards growth. The wetted front neither leads nor lags behind the dry. Further slowing finds the wetted portion leading. Throughout the process, a ***deceleration scarp*** (Fig. 2–27) is formed as the locus of contact between the wetted and the dry portions; this pinches out and disappears when the entire front has been wetted. Subcavitation hackle trails downstream from the scarp until velocities fall too low for its formation.

Quite another sort of scarp—the ***Sierra scarp***—is formed on deceleration in the absence of water at the free surface (Fig. 2–28). *The Sierra scarp results whenever an initially wet-running crack accelerates, cavitates, decelerates, and is overtaken by liquid water.* The sequence of events is sketched in Fig. 2–29. Once cavitation has occurred, the water does not remain static in the open crack but proceeds in the same direction as before under the thrust to occupy the narrowest part of the crack. It does not proceed with a straight front, however, but in a semiperiodically spaced set of fingers (the phenomenon is described as a Taylor instability). If in the meantime the crack front undergoes deceleration or arrest, the water may overtake it. It is the tips of the fingers which touch the moving crack front first; at each of the contact spots, the water aids crack advance (the velocity is well below the equivelocity point) and the crack front bows out slightly ahead of the dry portions. As cracking continues, water spreads laterally along the front, the locus of the wet boundary tracing a scarp—the Sierra scarp, named simply for its resemblance to a range of mountain peaks—on the crack surface.

Fig. 2–27. Deceleration scarp, crack running right to left. Timing marks, consisting of tertiary Wallner lines generated by pulses from a sonic generator, show successive positions of the crack front. Liquid water available at the free surface (top) intrudes into the decelerating crack front and at first retards it (upper right). × 100.

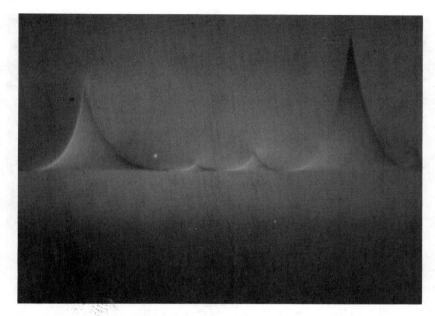

Fig. 2–28. Sierra scarp, formed where liquid water overtook a decelerating crack front. Crack ran from bottom to top. × 100.

In the typical occurrence, where air has not been completely stripped from the water before cavitation, subcavitation hackle appears on the downstream side of the Sierra scarp (Fig. 2–30).

Appearance of a Sierra scarp on a specimen is evidence that a sequence of acceleration-cavitation-deceleration has occurred with the presence of water at the crack origin and that there has been a corresponding history of stress rise and decay. It is most often met in wet thermal down-shock fractures, in which high tensile stress develops quickly at the chilled surface but falls off rapidly with distance into the hot specimen. Sometimes the crack may arrest, following development of the Sierra scarp, only to resume, cavitate, decelerate, and generate a new Sierra scarp. Indeed, such a sequence may be repeated several times before complete breakthrough is accomplished.

Many variations of the sequence can be observed. In some, the crack may come to a stop before the water overtakes it, and the Sierra scarp develops from an arrest line, only the wetted segments moving at first. In other cases, lateral spreading of water from the contact spots takes place too slowly to completely wet the crack front before a new rise in stress accelerates the crack to cavitate anew, leaving a series of closed ovals within which subcavitation hackle markings appear (Fig. 2–30). The possibilities seem to be endless, but all of them can be interpreted by patiently tracing their development across the crack surface.

Figure 2–31 shows an ***encounter scarp,*** generated when a slow crack, running dry, encountered a drop of water at the free surface. The water was drawn by capillarity into

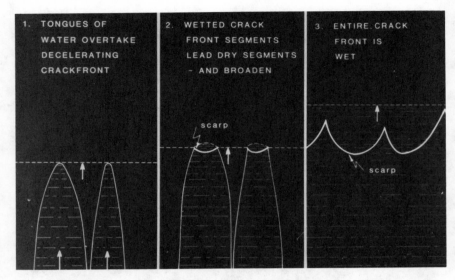

Fig. 2–29. Steps in the generation of a Sierra scarp. Fingers of water, overtaking the decelerating crack front, touch it first at a series of points. The wetted lengths of the front broaden as the crack runs, and they are advanced with respect to the neighboring dry lengths. The scarp is traced by the loci of contact of the wet with the dry regions. Eventually the dry regions are pinched out, completing the scarp, and the entire front is wet.

the crack front, reaching deeper and deeper as the crack advanced. The wetted portion of the front was aided by the water to advance beyond the position it would have reached if dry. A point of inflection in the crack front formed between the wet- and dry-running portions and a slight, abrupt change in level marks this point, the locus of which generated the scarp (Fig. 2–32). It runs from the point of first contact of the crack front with the surface water until the water had penetrated the full length of the crack front to the opposite side.

A *depletion scarp* can be formed when a slow-moving crack containing water in the near-front region runs in a specimen whose free surfaces are dry (Fig. 2–33). As a wetted crack proceeds, some of the water at its front combines with the free radicals released in bond breaking and is no longer available as liquid to follow the front. If the crack is leading at one side, and the local crack velocity is therefore greater in that region, the water may be used up there before it is exhausted in the more slowly moving portions (local velocities are meant in both cases; the crack front progresses at a single velocity[6]). Again a point of inflection arises between the wet and dry portions, and its locus becomes a scarp. It is distinguishable from the former case by the characteristic shape of any Wallner line that may intersect it; the Wallner line will show prominent leading at one edge, but in a region near that edge the advance is not as great as expected, and there will be a point of inflection between that portion and the leading one.

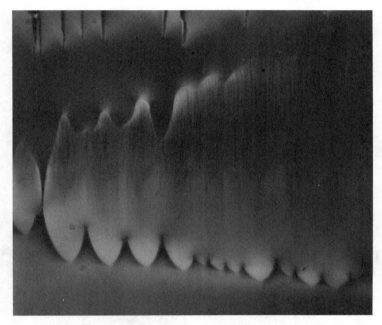

Fig. 2–30. A wet crack running bottom to top, having cavitated, decelerated to an arrest, was overtaken by water, and formed a Sierra scarp (bottom). It then accelerated, developed subcavitation hackle, and formed a cavitation scarp whose shape conforms to the location of the fingers of water which formed the Sierra scarp. At the left is a narrow neck where wetting was never accomplished.

A *Varner-Quackenbush scarp* or *Region II scarp* is the only scarp so far observed in which liquid-state water is not involved. When a crack grows slowly and accelerates gradually in a specimen in moist air, a record of the velocity of crack advance as a function of the crack-front stress, as that stress is calculated from conventional fracture mechanics, is shown in Fig. 2–34, together with the appearance of the crack surface and typical crack-front shapes along the crack path. The relationship is linear on the logarithmic plot at first; this is Wiederhorn's Region I.[7] Here the crack front is slightly bowed, the convex side pointing in the direction of crack propagation, as moisture is able to reach all parts of the front. (The slight retardation of the edges with respect to the center is caused by Poisson relaxation and consequent partial stress relief at the free surface.)

With increasing velocity, the middle begins to lag, showing that the concentration of moisture is less there than at the edges, where moisture from the atmosphere has a shorter distance to diffuse along the front, and the initial linearity of the curve begins to break down slightly.

A velocity is then reached where no moisture whatever reaches the middle of the crack front, and Region II is said to begin. At first the dry portion of the front is confined

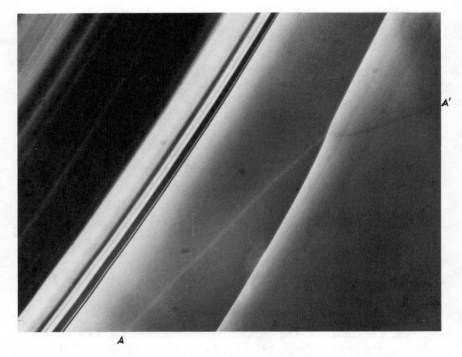

A'

A

Fig. 2–31. An encounter scarp (running from A to A') formed as the crack, running slowly left to right, encountered water at the bottom edge. Note that the lower end of the Wallner line (light line at the right) is advanced in the wetted region. × 100.

to the middle, but with increasing velocity a growing fraction of the front becomes dry and the moist outer strips become narrower. Their final disappearance marks the end of Region II and the beginning of Region III, where atmospheric effects on crack spread are absent. Between the moist and dry crack-front segments there is a point of inflection in the crack front, and its locus becomes the Varner-Quackenbush scarp, named for those who first observed it and explained its significance.[8] In practice this scarp is seldom seen, and then only as a parenthesis-shaped mark whose concave side points in the direction of crack travel. (Note that this is the reverse of the general rule for rib marks.)

A Varner-Quackenbush scarp also forms in a crack decelerating in a moist atmosphere. Its sense of curvature is then reversed with respect to the directions of crack spread.

2.8   Parabolic Markings

Kies figures[9] are formed when new cracks originate independently in the stress field in front of a fast-running crack. A circular crack radiates from such a new origin until it

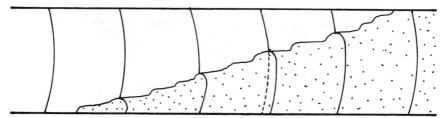

Fig. 2–32. Generation of an encounter scarp. The crack running from left to right encounters liquid water at the lower free surface. The wetted portion of the crack front (stippled) advances with respect to the dry and a scarp forms along the boundary between them, progressing through the thickness until the entire front is wet.

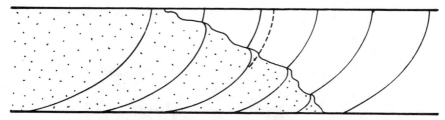

Fig. 2–33. Generation of a depletion scarp. The advanced side of the crack front, running wet, is depleted of water before the slower side and is retarded behind its expected position; the boundary between wet and dry portions of the crack front is traced as a scarp on the crack surface.

encounters the advancing main front; this point of interception becomes the nose of a parabola whose sides are then traced by the locus of intersection of the two moving fronts (Fig. 2–35). The focus of the parabola, that is, the point of origin of the new crack, can be recognized as the center from which hackle "streamers" radiate in all directions.

Kies figures are to be seen on crack surfaces of such materials as PMMA (polymethyl methacrylate, or Plexiglas*), cellulose acetate, fine-grained steel, and single-crystal KBr. They have also been observed in an inclusion-filled obsidian.[10] They might also be expected in ceramics which have potential origins scattered throughout their bulk, but it has not been reported. They have been discussed here because of their possible relevance to high-temperature failure of brittle materials containing internal centers of weakness.

2.9   Spurious Markings

Markings on fracture surfaces which have nothing to do with fracture may sometimes resemble the classic marks. In pressed glassware, for example, cords may be folded to

---

*Neville Chemical Co., Pittsburgh, Pa.

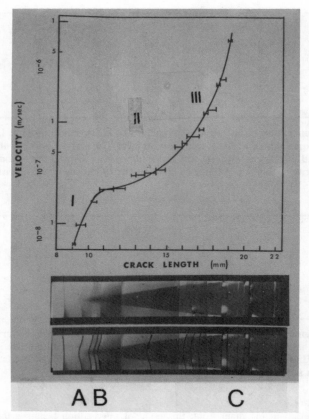

Fig. 2–34. Varner-Quackenbush scarp (wedge-shaped dark line separating light and dark areas) in a mosaic photo and marked to accentuate successive positions of the crack front. The corresponding velocity curve is plotted above to the same scale of crack length. Through Region I the entire crack front is affected uniformly by water from the environment and advances with normal slight lag at the edges, a consequence of Poisson relaxation there. At A, the crack middle lags as its access to moisture is curtailed by the increased velocity. At B, the middle has become completely dry and this dry region broadens, outlining the scarp, as velocity increases until the entire crack front has become dry at C and crack growth is then independent of environment.

yield lines resembling Wallner lines; their true significance will be evident only through their inconsistency with the other, fracture-generated marks.

Polycrystalline materials may also develop spurious markings as a consequence of their history of fabrication. For example, auger laminations may easily lead to false interpretation if other markings are not taken into account (Fig. 2–36).

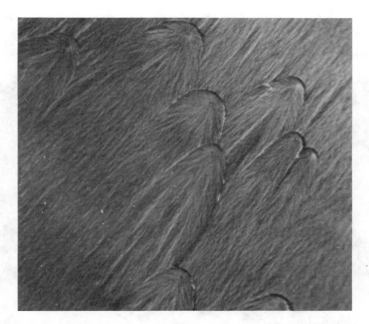

Fig. 2–35. Kies figures (parabolic markings) on a polymethyl methacrylate (PMMA) crack surface. Crack direction upper right to lower left. ×500.

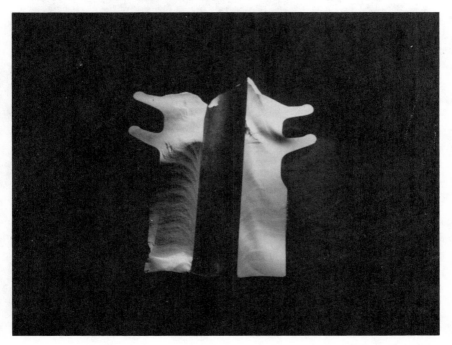

Fig. 2–36. High-tension porcelain insulator broken axially shows lines from auger laminations during forming. In the right half can be seen broad primary Wallner lines originating from contour details, with some twist hackle.

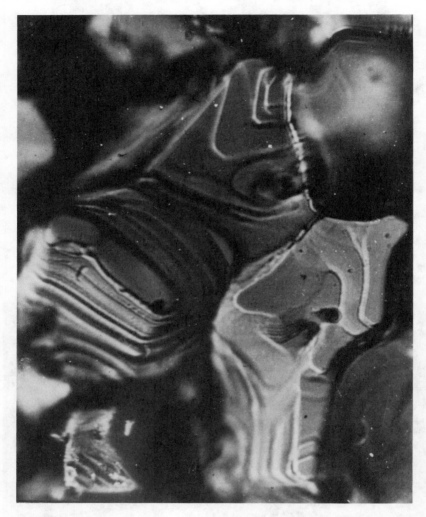

Fig. 2–37. Etch figures in grains of a BeO ceramic broken at 1700°C in a moist atmosphere.

In some cases, growth terraces or etch pits may yield puzzling details on individual grains of polycrystalline ceramics broken at high temperatures (Fig. 2–37). In such cases, the generation of closed figures by the lines may identify their nature.

3.   The Pattern of Forking

3.1   Angle of Forking

A crack running under increasing stress reaches an effective limiting velocity, or terminal velocity, and then may split into two cracks. The process is called *velocity forking, velocity branching,* or *velocity bifurcation.* The angle subtended by the emerging crack pair in a thin specimen is dependent on the ratio of the two principal stresses in the plane of the free surface. Preston[1] first called attention to this dependence and listed the principal cases. If a thin sheet, supported at its edges, is subjected to point pressure at its center, its back surface is brought into equal tension in all directions, that is, the ratio of the principal stresses, σ_y/σ_x, has the value of $+1$. The maximum angle of forking in this case is 180°. If a thin lath is broken in cross bending or in simple tension, there is stress in one direction only, the stress ratio is zero, and the maximum angle of forking (this must always be measured close to the point of forking) is 45°. A cylinder burst by internal pressure has a hoop tension which is twice as great as the axial tension; the stress ratio is one-half, and its maximum angle of forking is 90°. And finally, if a thin-walled tube is broken by torsion about its axis, a tensile stress develops at an angle of 45° to the axis, and a compressive stress of equal magnitude develops perpendicular to it. The ratio is therefore -1 and the angle of forking is 15°. Figure 3–1 summarizes the results. Intermediate values of the forking angle can be observed in more complex geometries and in more elaborate systems of stresses.

3.3   Meander Cracks

A crack whose course resembles that of a meander river is especially characteristic of center-heated plates, such as window panes heated in the center by the sun's radiation, their edges being kept cool (Fig. 3–2). The meander crack is not limited to center heating, however, and so cannot be used as proof positive that a failure resulted from this condition. It may also appear in failures under violent impact, in the later stages of breakup.

No successful analysis has been applied to this rhythmic cracking pattern, but it is clearly not random, as the typical symmetry of the undulations in velocity-forked meander cracks plainly demonstrates (Fig. 3–3).

3.4   Intersecting Cracks

For a variety of reasons it may be necessary to determine which of two crack systems in a specimen was first to occur. It may be a case of learning which of two events caused a failure and which occurred at a later time, as when fragments from a failure have been carelessly handled and further fragmented. Or it may be a question of finding the primary origin of a failure in a specimen in which there are several independent crack origins, all of them the result of a single event.

In every case, the principles are the same. If a particular crack runs into another, preexisting crack, it is arrested. Running under the influence of tensile stress concentrated

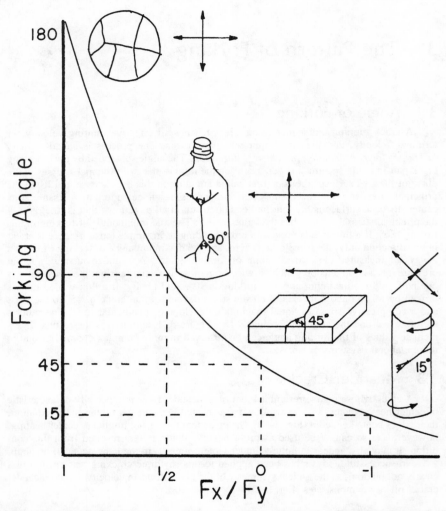

Fig. 3–1. Relation between the maximum angle of velocity forking and the ratio between the principal stresses. After Preston (Ref. 1).

at its sharp tip, it is abruptly blunted by encounter with a preexisting crack, and it stops. If the stress system is still active, a new crack may begin on the other side, subject to the usual requirement for the presence of a flaw large enough to grow under the active stress. It is practically inevitable that such a flaw will not be found immediately opposite the site of the arrested crack, but at some distance from the point of intersection. The later crack jogs, therefore, before resuming on the other side of the earlier crack.

Fig. 3–2. Meandering crack path in a plate broken by heating the center, the edges being cold. Crack ran from right to left.

Even if the jog is very short, the fact of renucleation will be evident on the crack surface, since the new segment of the later crack will develop from a surface flaw, and this fact will be revealed from such signs as Wallner line shapes and twist hackle configuration. By contrast, the earlier crack will show no discontinuities in its own surface in the region of the intersection.

For example, when failure under impact generates a family of radiating cracks, the slender fragments are brought under cantilever stress by the still-acting external force. The edges of the fragments, if not formed at velocities in the terminal range where they are roughened by mist hackle, may offer no stress-raising flaws to initiate cantilever cracking. Cracking may then begin at a flaw in the free surface, that is, at an independent origin. Observation that the cross-break cracks terminate in the radiating cracks is then sufficient to identify them as later in the breakup event.

Under special circumstances, cracks may cross one another without jogging. This happens when one, strongly leading at the front surface but not yet broken through to the back surface, is intersected by another, leading at the back surface and not yet through to the front. Both crack surfaces show discontinuity at the intersection (Fig. 3–4). This circumstance arises in plates broken by twisting, and in specimens broken by impact where radial and circumferential cracks develop almost simultaneously.

3.5 Phantom Cracks

Under the proper conditions, cracking is a partially reversible phenomenon. Provided that the crack surfaces have not become contaminated with sorbates or particulates and

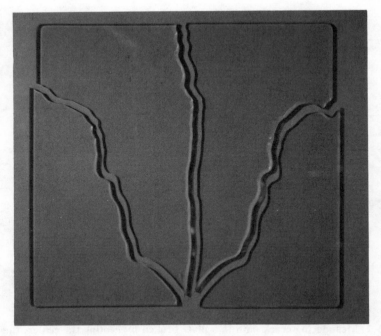

Fig. 3–3. Symmetrical meandering of the velocity branches of a crack in a glass sheet broken by central heating.

that they are still in register, healing may occur to the extent that 90% or more of the original strength is recovered. The precise conditions have not been explored, but apparently recovery depends in part on access to moisture.[2] Too moist a condition is not favorable, nor may the surfaces be too dry. The practical result is that sometimes a crack may be seen that seems to be isolated, having no connection with other cracks in the specimen and having no point of origin within its length. This is most often encountered in the midst of a system of cracks radiating from an impact center, and then it is not difficult to recognize. It may also be seen in circumstances that are not so easy to interpret, but if the healing possibility is remembered, the proper origin of such cracks can be located in the open portion of the earlier crack stage, where healing has not occurred.

3.6   Crack-Surface Profile

The profile of the crack surface, that is, its degree of lateral curvature, and its angular relationship to the free surfaces indicate features of the stress system acting at its front.

In a sheet subjected to pure in-plane tension, a crack develops a flat surface which is perpendicular to the free surfaces. The condition is met in certain thermal breaks and near the origin in a bursting cylinder, for example.

Under pure bending stress, one surface is initially under tension while the opposite

(A)

(B)

Fig. 3–4. (*A*) Primary crack surface intersected by a secondary crack, seen edge-on as a vertical line at the center of the photo. The mist hackle, secondary Wallner lines, and trailing-edge hackle of the primary crack show no indication of interference from the secondary crack which must be concluded to have occurred after the primary crack was complete. (*B*) Crack surface of one of a pair of intersecting cracks. At its leading edge (bottom) the mist hackle and secondary Wallner lines are not disturbed by the intersecting crack, but in the upper half of the photo a set of primary Wallner lines and twist hackle have been generated, showing that the second crack was present before the one whose surface is shown was completely through the thickness.

Fig. 3–5. Cantilever curl shown by the profile view of a crack formed in cross bending. The crack originated in the lower surface, first running straight upward, then curved to the left.

side is under balancing compressive stress and there is a line of zero stress (the neutral fiber) down the middle (Fig. 3–5). The resulting crack is flat and perpendicular to the free surfaces along the tensile side, where the crack front leads prominently. But the initial stress system is altered by relaxation of stress at the tensile side as the crack front passes. The neutral fiber position is then shifted and the tensile stress field at the crack edge becomes so strongly curved that the crack front sees a principal tension whose axis is ill-defined. Consequently the crack surface curves off, or "curls" as its trailing edge reaches the initially compressive region to complete the separation. (It is this same shift in stresses that generates twist hackle at the trailing edge of cracks formed in bending.) The curvature of the crack surface under bending, or cantilever stress, is called ***cantilever curl***.

Cantilever curl may also be observed on the crack-generated surface, where it forms a ridge about two-thirds of the way across the specimen. The ridge-axis perpendicular lies in the plane of stress application, and so, in a perfectly homogeneous specimen, it points to the position of the origin (Fig. 3–6(A)). In cases where the origin is found to be offset from its expected position (Fig. 3–6(B)) it must be inferred that the specimen failed as the result of an unusually severe flaw. In bending strength tests on cylindrical rods this indicates that failure has occurred at stresses below the values indicated by routine calculations.

(A)

(B)

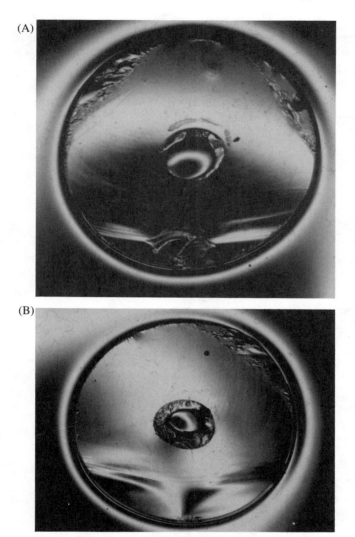

Fig. 3–6. Crack surface of a glass capillary tube broken in cross bending. (*A*) Origin (top center) lies on the normal to the ridge of cantilever curl. (*B*) Origin is offset to left of the normal, and the mist-hackle bordering the fracture mirror is displaced.

The strength of a rod specimen in three-point bending, with an origin displaced lengthwise a distance x and rotated through an angle ϕ from the point of expected maximum tension, can be computed[3] from the equation

$$\sigma_B = 8/\Pi \cdot (L-2x)/D^3 \cdot K \cos\phi$$

where σ_B is the breaking strength, L the distance between supports, D the rod diameter, and K the load at failure.

4. The Seeds of Failure

The flaw from which cracking originated is often the prime focus of attention, particularly if failure has occurred under relatively low stresses. Its nature may give clues to its history and to the kind of event which was responsible for its presence. A dangerous flaw may be formed during almost any stage of the manufacturing process, as, for example, from contaminants in the raw materials; improper formulation, mixing, and processing; proof testing; and handling in inspection and packaging. Damage can occur during shipping, warehousing, and distribution. And certainly it can result from wear and abuse in service. The characteristics of some typical flaws will be described, but the list will be incomplete. Almost every product will have its own peculiar susceptibilities with respect to flawing.

4.1 Hertzian Cracks

Localized impact by or against a hard, relatively smooth, blunt object can form a **Hertzian cone** (it is more properly called a Hertzian conoid), constituting an entire failure, as when a BB shot strikes plate glass and pops out a conoid (Fig. 4–1), requiring replacement of the pane. More often the Hertzian conoid does not develop completely, and its beginnings, such as the initial **ring cracks,** while often invisible, remain as flaws from which failure may develop at some later time under some other form of loading.

The entire life history of Hertzian conoid development will be presented in this section, because in practice it may proceed through any number of its stages, and then arrest, constituting a potential crack origin. Usually not all of the partial Hertzian is effective in initiating cracking at the time of failure under later loading, and it is therefore important in failure analysis to be able to recognize it as Hertzian from only a portion of its crack surfaces.

Consider a hard sphere striking a flat plate. Under the pressure of impact, the plate is forced to conform to the shape of the sphere (the sphere also deforms). As the process continues, the surface of the plate is stretched, the circle of contact broadens, radial tension at its circumference increases more and more, and cracking begins. First, a shallow circular crack forms at the surface (Fig. 4–2), sometimes multiple concentric cracks. If the impact is light, the damage may be limited to these checks, which persist as surface flaws.

If the impact is more severe, the ring-shaped check at the surface penetrates deeper, flaring outward in the form of a conoid in response to the compressive stress operating axially. A collar of mist hackle forms and then the surface develops smoothly (but typically marked with tertiary Wallner lines) and the conoid vents to the back surface.

If the impact is sufficiently violent, a radial crack may be initiated in the rough details of the mist hackle. It runs faster than the conoidal crack and bisects it and, together with other radial cracks which may then form, it runs radially out from the center of impact into the surrounding material, leading at the back surface.

The impacting body, if its energy is sufficient, continues to exert inward force and

Fig. 4–1. Hertzian cone (percussion cone) caused by high-velocity impact from a small, hard body against the center.

may break off the slender radial fragments by means of circumferential cracks which lead at the front surface and show the cantilever curl described above.

4.2   Sharp Particle Impact

While the Hertzian conoid formed by impact from a smoothly rounded body can be described reasonably simply, a ***peck,*** or indentation[1,2] from a sharp, hard, usually irregular agent cannot. Such "pecked" damage sites commonly include loss of a small amount of material by pulverizing. They also include checks of irregular shape, some roughly perpendicular to the free surface and some nearly parallel to it (Fig. 4–3).

4.3   Bruises

Contact with a hard, rough, blunt object can result in a ***bruise,*** a sort of damage site which is ill-defined. Crushing is the distinctive feature, usually associated with checks extending from the impact site laterally and in depth. Like the peck, it includes pulverizing of the surface, together with microcracking or checking, but unlike the peck may not involve any loss of material.

(A)

(B)

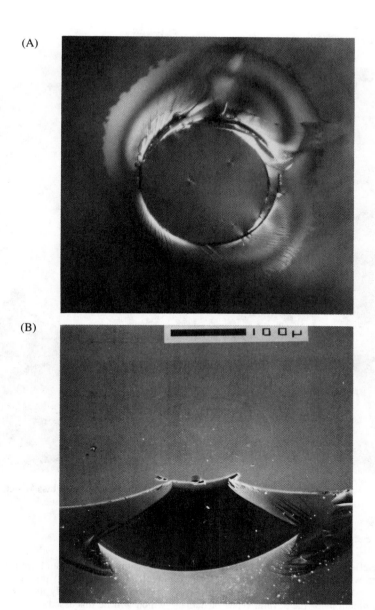

Fig. 4–2. Initial stage in the formation of a Hertzian cone. (A) Appearance at the free surface, (B) cross-sectional view. The shallow circular cracks formed first, then began to flare outward, generating the beginnings of a conical crack surface. The area of contact with the projectile is undamaged. ×75. Courtesy of J. R. Varner.

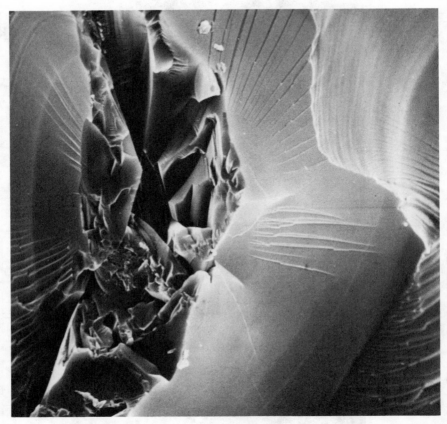

Fig. 4–3. Scanning electron micrograph of complicated "peck" on glass which has been struck by a grain of sand. × 1,000. Courtesy of J. R. Varner.

4.4   Scratches and Gouges

The effect of drawing a piece of hard material across the specimen can have a range of effects which depend on the relative elastic moduli, shapes, contact pressure, and surface condition. The result is a string of damage centers which typically include complex combinations of the several types of damage which will be discussed.

A fixed-abrasive point drawn slowly across a brittle material under light pressure can leave a shallow groove in the surface, or it may remove turnings that are reminiscent of metal turnings from a machining operation. Indeed, the principle is being explored as a possible process for generating optical-glass lens surfaces by single-point diamond machining. Figure 4–4 shows the groove left in the specimen, together with a length of the

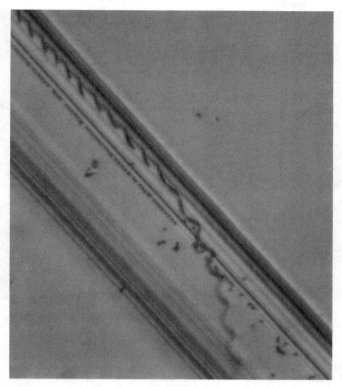

Fig. 4–4. Groove left by scratching glass with a diamond tool, together with a turning which separated from the specimen after passage of the diamond. ×500.

turning removed. The turning separates from the specimen immediately behind the tool. Its tight curl is the consequence of permanent densification of the glass under high local loading under the tool, an effect which falls off with depth below the surface. The resulting strain gradient causes both spontaneous separation of the turning from the surface and tight curling of the separated strip. Under heavier pressure, "butter-fly" checks are developed at the groove edge and chips fly off on either side. With sufficient applied force, the tool ploughs families of such chips and leaves a deep gouge. An irregularly shaped tool may show all of these stages along the gouge path (Fig. 4–5).

A loose particle, rolled across a surface under load, leaves a damage track that can include indentations, scratches, Hertzian ring checks, and gouges. The checks which are members of this track extend to a depth between four and five times the particle diameter, an important consideration when specimens are to be finished by loose-abrasive grinding

Fig. 4–5. Gouge mark left by an irregularly shaped tool under moderate pressure. Crossed Nicols; ×75.

and polishing, high strength being important to the performance of the product. It shows how much stock must be taken off at a particular stage to remove the damage introduced in the preceding one. In failure analysis it is a useful rule of thumb to estimate the size of the particle responsible for the origin flaw.

4.5   Chatter Marks

When surfaces come into sliding contact they may tend to seize one another and thus exert tensile stresses in the direction of sliding. *Chatter marks* result, a series of checks which originate along the line of contact and run initially perpendicular to it, then curve off at the edges to show concavity in the direction toward which contact progressed. (Note that this is opposite to the general rule for curvature of rib marks on the crack surface.) Figure 4–6 shows a typical chatter mark.

4.6   Chill Checks

Sudden local chilling of a hot specimen contracts its surface, develops radial tensile stress, and can form a *chill check* (Fig. 4–7). A large temperature difference between specimen and chilling medium, flaws in the specimen surface, and a high specimen thermal expansion coefficient all contribute to the likelihood of chill checks. The depth to which they will penetrate below the surface depends on such conditions as the tempera-

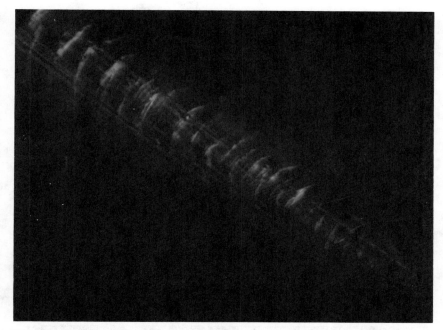

Fig. 4–6. Chatter mark from a tool traveling from left to right. Crossed Nicols; ×120.

tures, area and intimacy of contact, and duration of contact, as well as the thermal properties of the specimen and the chilling medium.

Chill checks are hard to characterize in any general way, because the chill is so often accompanied by mechanical effects at contact, which influence the course of their development.

4.7   Other Potential Origins

Recognition of frequently encountered origin types or locations can be used to correct errors in all stages of design, manufacture, and service. The preceding sections have emphasized those flaws which are formed by mechanical or thermal abuse, because such flaws are themselves products of cracking processes. There are other sorts of potential crack origins. Product design may furnish stress-raising features, such as sharp changes in contour, especially at inside radii. Identification of these as systematic origins of failure in test or service provides the key to redesigning for improved performance.

Impure or contaminated materials may introduce inclusions, with attendant stress concentrations and even microcracks. Fabrication may introduce dangerous reentrant angles as, for example, lapped-over baffle marks, shear marks, and mold parting lines in machine-blown glassware. External agencies may etch or pit the surface chemically or may raise blisters.

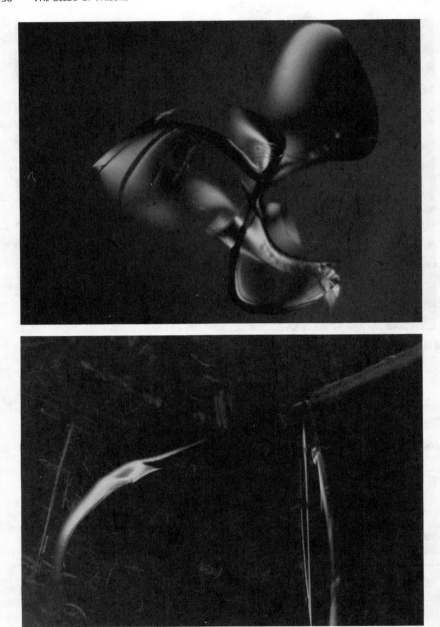

Fig. 4–7. Chill check produced by heating glassware at 450°C and quenching with a copper sphere at room temperature. Subcrossed Nicols; ×50.

Such potential origins may or may not be evident prior to failure. Fractographic analysis of failed specimens must include examination of the origin region for indications as to what agency was responsible. Detailed knowledge of the raw materials, processes of manufacture, and conditions of service may be necessary to identify the cause exactly, but precise description of the origin morphology by the fractographer may be sufficient to enable an expert in the technology of manufacture and use to recognize it.[3-10]

5. Estimation of Stress at Failure

The magnitude of stress concentrated at the origin at the moment of a failure is pivotal in deciding whether a specimen was unduly weak or whether excessive stress was applied in service. Fortunately, there are several indications of the failure stress that are available for use in postmortem examination. None of them provides an exact value, but the approximations are nevertheless useful.

5.1 The Origin Flaw Size

As noted, the stress necessary to run a through crack in a sheet of glass under simple tension is inversely proportional to the square root of the crack length; thus, after calibration of the proportionality constant, a measure of an origin crack length in postmortem examination of a failure can be used to infer the stress at failure. It is not as simple as that, however. In the first place, the proportionality holds only if the calibration origin cracks have the same history as the flaw in the failure. As Shand[1] has shown, the strength of glass increases with the time of unstressed aging in air. One hour from the time of flawing, the strength of his specimens had increased to 116% that of specimens tested immediately; in a day they had increased to 122%, in a week to 124%, and in a month to 130% and still rising. Thus we have the seeming paradox that specimens tested in moist air are weaker than when tested dry, yet storing specimens in moist air yields strengths which increase with the time of storage. Immersion in water during aging was still more effective; after 36 hours in water the dry strength had increased to 150% that of unaged specimens. Heating also had a profound effect on the strength of flawed specimens. Only 12 minutes at 520°, that is, below the annealing temperature, and cooling, raised the strength to 140% of that of the controls.

Neither the unstressed aging or heating treatments had changed the geometry of the flaws as they could be seen under the microscope, but apparently had blunted the flaw tips. Even faster blunting has been reported[2] for aging under loads 75% of the threshold value for crack running.

The length of time in which the specimen has been under load has a pronounced effect on the load which can be supported without failure. Shand[3] showed that, for a few seconds, a specimen can support twice the load that would break it in ten days. This effect is called static fatigue, and is a consequence of the fact that flaws under sufficient loading can grow slowly, but at increasing rates as they grow larger, until the growth rate is so fast as to be termed catastrophic, that is, it constitutes failure. The effect is limited, however, and no case of such delayed failure of soda-lime glass has been reported after 11 days, in agreement with calculations which equate the threshold crack velocity with the rate of crack-tip blunting.[4]

This serves as a warning of large errors to be anticipated in estimating failure stresses from examination of flaw dimensions, unless the history of the specimen is known and unless the effects of that history can be taken into account.

The effective dimensions of some origin flaws can be measured with the microscope

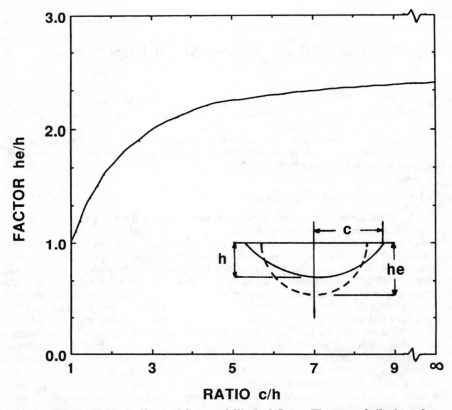

Fig. 5–1. Stress-raising effects of flat semielliptical flaws. The term h_e/h gives the equivalent radius h_e of the flaw. After Shand (Ref. 1).

on the crack-generated surface, using Wallner lines and hackle to identify that portion of the visible flaw which participated in nucleating the crack. These dimensions can then be normalized[1] to give the equivalent radius of a semicircular crack (Fig. 5–1), and used to calculate the stress required to extend it under the relevant conditions.

5.2   The Mirror Radius

In a high-stress failure, the crack accelerates from the origin flaw until it approaches terminal velocity, and its relatively smooth surface becomes successively more and more roughened by mist hackle until it undergoes velocity forking. The *mirror region* is the smooth portion of the crack surface bounded by mist hackle. The term is not to be used to describe a crack surface that is merely smooth. Whether the mirror region is circular or not, the mirror radius is taken as the distance from the origin, measured along the free surface (or a defined distance in from the specimen surface) (Fig. 5–1) to some recog-

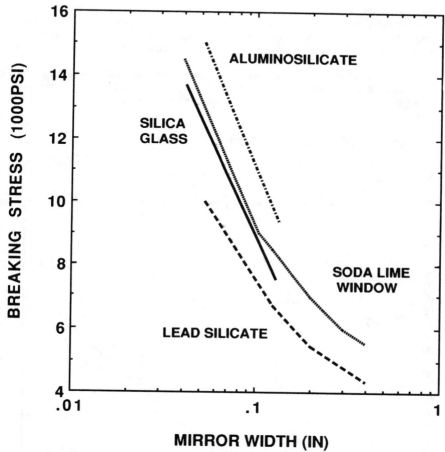

Fig. 5–2. Strength of glasses as a function of R, the mirror radius. From Shand (Ref. 1).

nizable feature such as the first detectable mist hackle. There are, accordingly, several mirror radii that can be used, and these are designated R_{mist}, R_{hackle} (this is measured to the beginning of the coarser details in the mist hackle), and R_{fork}.

Still another measure of the mirror radius involves plotting the cumulative number of microcracks exposed at the free surface as a function of distance from the origin and extrapolating the plot back to unity.[5] Long before velocity forking occurs, the tendency to fork is evident, as already noted in connection with mist hackle.

The relationship generally used to relate mirror radius to stress at the origin at failure is simply:

$$\sigma_F = a/\sqrt{R} + b$$

where the constants *a* and *b* are determined by the material, the specimen geometry, the particular radius measured, the stress system, and the test temperature.[6-9] Particular examples will be given in the sections dealing with typical failures and specific materials. The equation is not applicable to low-stress failures, in which the mirror radius is large or nonexistent.

The mirror radius is generally more satisfactory for estimation of failure stress than is measurement of the origin flaw, because it is not so much influenced by specimen history, nor by environmental species. But stored stresses, as in tempered glass, will affect it.

5.3   Extent of Fragmentation

It is common experience that the more violent the event the greater the number of fragments produced. A linear relationship has been observed for bottles burst by internal

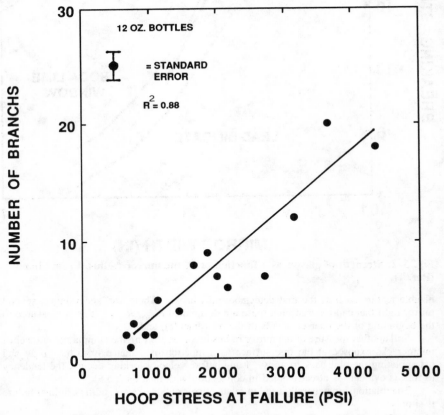

Fig. 5–3. Fragmentation of 12-fl. oz. soft-drink bottle as affected by bursting pressure. After Frechette and Michalske (Ref. 10).

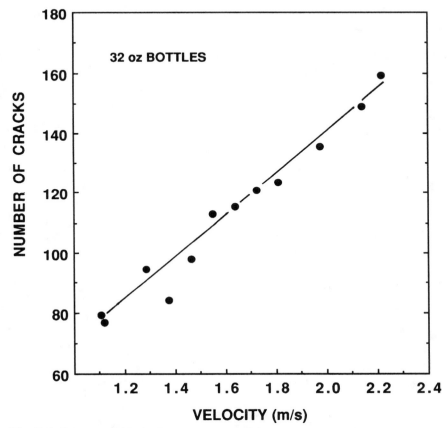

Fig. 5–4. Fragmentation in the central zone of a beer bottle as affected by the velocity of impact. After Frechette and Yates (Ref. 11).

pressure[10] (Fig. 5–3) and for bottles broken by impact against the sidewall[11] (Fig. 5–4). The usefulness of these relationships is somewhat limited in practice, however, by the necessity to evaluate the proportionality constants for different bottle designs and conditions of test.

6. Effects at Inclusions

Formation of wake hackle and gull wings on the crack surface when the crack front traverses past an inclusion has already been discussed. In many cases they may be the most useful of all observations in establishing the path and direction of cracking. Additional effects at inclusions have to do with the stress fields and the microcracks that may be associated with them. Besides their use in failure analysis, they are useful in helping to recognize the nature of crystalline contaminants ("stones") in glass and in ceramic materials.

6.1 Spontaneous Cracking

The effect of an inclusion on cracking is to be understood in terms of the stress fields involved. The inclusion which has undergone more shrinkage than the matrix is subjected to isostatic tension throughout. At the same time, the matrix is brought under circumferential compression and radial tension whose magnitudes decrease with the square of the distance from the inclusion surface at a rate which is inversely related to the inclusion diameter. The converse obtains in the case of an inclusion which has shrunk less than the matrix and is consequently under isostatic compression, while the matrix is in circumferential tension and radial compression. Where these stresses are sufficiently high, spontaneous cracking occurs, as shown in Fig. 6–1. The inclusion may be surrounded by cracks running outward into the matrix (inclusion under compression) or the inclusion itself may be cracked and encircled by a crack (inclusion under tension). These are the microcracks[1-3] which are characteristic of certain ceramics such as the conventional quartz-type porcelains. They provide potential origins that can lead to catastrophic failure under applied stress, or in some special cases they can provide elasticity to the system and lessen the likelihood of failure.

6.2 Critical Inclusion Size

Not all inclusions should be listed as dangerous defects. If their dimensional changes relative to the matrix are small, the stresses in and about them will be insufficient to generate microcracks, and if they are remote from the sites where service stresses peak, they may have little or no tendency to reduce service performance. Their deleterious effect does depend on size, however. It is well known that, although a wire of small diameter can be safely sealed to provide an electrical lead through a glass envelope such as the base of an electric light bulb, a wire of large diameter cannot. Similarly, an inclusion of small diameter may be innocuous, whereas a larger one may lead to failure.

The through-glass wire, with its lower coefficient of thermal expansion than the glass, will shrink less on cooling from the sealing temperature and will cause circumferential tension in the surrounding glass. The maximum tension will occur at the interface with the wire. The magnitude of the maximum tension does not depend on the diameter

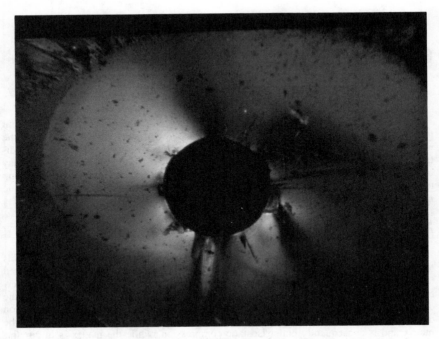

Fig. 6–1. Chromite inclusion (black disk at the center) at origin of fracture in green bottle glass showing stain field and radiating cracks. Crossed Nicols; × 65.

of the wire, but the rate of decay of stress with distance from the interface does depend on diameter, as shown in Fig. 6–2.

A flaw capable of developing into a crack can be expected to be located at the interface, and so its critical dimension must be measured from there. The Griffith equation shows that the farther a flaw extends from the interface, the less is the stress necessary to propagate it, as Fig. 6–2 indicates. If the stress field about the inclusion decays rapidly with distance, as it does in the case of the 0.1-mm wire, the tensile stress field is everywhere too low to run a flaw which begins at the interface. On the other hand, the tensile stress around a 1.0-mm wire falls off much more gradually, so that a point is reached (point A in Fig. 6–2) where the tensile stress field does become sufficient to propagate a flaw reaching outward from the wire surface; the flaw therefore grows larger. Crack extension continues outward until a second point is reached (point B) where the tension is no longer sufficient to extend a flaw of the size to which it has then grown, and the crack stops. (As a practical matter it may already have grown large enough to provide a leak to air; it may also serve to initiate failure under subsequent thermal or mechanical stress.)

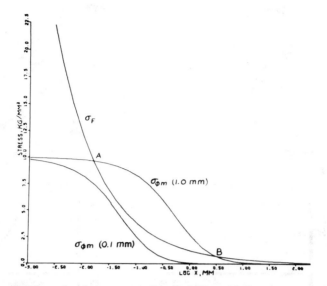

Fig. 6–2. Tensile stress in glass surrounding inclusions of diameter 1.0 and 0.1 mm, respectively, together with a plot of the stress necessary to extend a crack of length *x* measured from the inclusion surface. A 1-mm inclusion will run cracks of length A to length B. See Oel (Ref. 4).

6.3   Crack Deviation at Inclusions

Figure 6–3 shows the effect of stress fields, in the region of an inclusion, on the course of a running crack. The crack tends to deviate locally from its plane to avoid intersecting an inclusion that is under tensile stress relative to the matrix, and it deviates in such a way as to intersect an inclusion that is under compression.

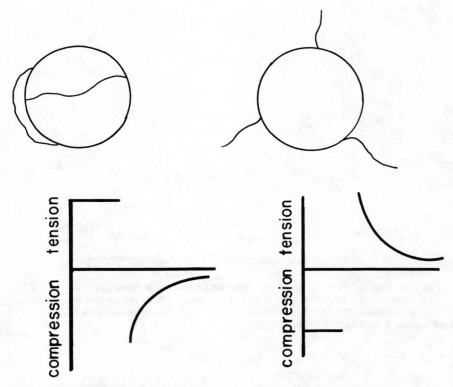

Fig. 6–3. Cracks associated with an inclusion under tensile stress (left) and compressive stress (right) together with the corresponding stress distributions.

7. Anisotropic Materials

The general phenomena of cracking have been discussed as they occur in materials which have no directional properties as far as fracture is concerned. The same phenomena that occur in glass fracture are to be expected in anisotropic media also; however, a few, such as the scarps, have not yet been observed.

Single crystals of certain kinds, for example, quartz, behave as glass does in cracking, although the stress fields developed in them may be modified somewhat by anisotropy in their elastic properties. Thermal stresses may be altered markedly from those in glass by reason of anisotropy in thermal expansion and thermal diffusivity. Such effects should be considered as possibly important in specialized cases, but they will not be treated further here.

7.1 Cleavage and Parting

Many single crystals, of which mica is the classic example, are weakly bonded across certain planes. Cracking occurs more easily along planes, and is then called *cleavage*. (But metallurgists sometimes use the term cleavage to mean transgranular fracture.[1]) Cleavages must obey the laws of crystal symmetry. Thus, if there is cleavage parallel to one of the crystallographic axial planes in a cubic crystal, there must be cleavage parallel to each of the other axial planes also, since the three axial planes are symmetrically equivalent to one another in cubic crystals. This particular cleavage is called cubic because the three planes are parallel to the faces of the cube, and is characteristic of NaCl (rock salt), MgO (periclase), and CaO. If in another cubic crystal there is a cleavage plane perpendicular to a body diagonal, symmetry requires a total of four such planes, because the four body diagonals are symmetrically equivalent. This is octahedral cleavage, typified by CaF_2, UO_2, and CeO_2. In other crystal systems, analogous sets of cleavage planes exist, as summarized in Table I.

The existence of a cleavage means that, other things being equal, cracking will proceed along the cleavage plane. But cracking is by no means limited to the cleavage. The extent to which cracking does follow cleavages is described only roughly by such terms as perfect, good, fair, or difficult. Even in crystals with "perfect" cleavage, a crack may develop what mineralogists call *conchoidal fracture,* that is, forming fracture surfaces fancied to resemble certain seashells, the axis of principal tension being too severely misaligned with respect to cleavage planes for cleavage to play a major role in the fracture. Then the cracking of the crystal proceeds just as it does in glass.

The cleavages ascribed to the natural and artificial minerals by mineralogists working with hand specimens are usually accurate, but additional cleavages may be observable by microscopic examination. Wolff and Broder[2] described an elegant technique for defining microcleavages, but it does not appear to have been widely exploited. Because of its great potential for contributing to the fractography of single crystals and polycrystalline materials, it will be described briefly. A single crystal is cut or ground to be roughly spherical. It is then rolled under moderate pressure between two sheets of very coarse SiC abrasive

Table I.   Cleavages Characteristic of the Crystal Systems

System	Name	Form	Cleavage Planes
Isometric	Cubic	{100}	(100)(010)(001)
	Octahedral	{111}	$(111)(1\bar{1}1)(\bar{1}11)(\bar{1}\bar{1}1)$
	Dodecahedral	{110}	$(110)(1\bar{1}0)(101)(10\bar{1})(011)(0\bar{1}1)$
Tetragonal	Basal	{001}	(001)
	Prismatic*	{110}	$(110)(1\bar{1}0)$
	Pyramidal*	{111}	$(111)(1\bar{1}1)(1\bar{1}1)(\bar{1}\bar{1}1)$
Hexagonal	Basal	{0001}	(0001)
	Prismatic*	{10$\bar{1}$0}	$(10\bar{1}0)(01\bar{1}0)(\bar{1}100)$
	Pyramidal*	{11$\bar{2}$0}	$(11\bar{2}0)(2\bar{1}\bar{1}0)(1\bar{2}10)$
	Rhombohedral*	{10$\bar{1}$1}	$(10\bar{1}1)(1\bar{1}01)(01\bar{1}1)$
Orthorhombic	Basal	{001}	(001)
	Pinacoidal	{010}	(010)
	Pinacoidal	{100}	(100)
	Prismatic	{110}	$(110)(1\bar{1}0)$
	Prismatic	{101}	$(101)(10\bar{1})$
	Prismatic	{011}	$(011)(0\bar{1}1)$
	Pyramidal*	{111}	$(111)(1\bar{1}1)(\bar{1}11)(\bar{1}\bar{1}1)$
Monoclinic	Basal	{001}	(001)
	Pinacoidal	{h01}	(ho1)
	Prismatic	{110}	$(110)(1\bar{1}0)$
	Prismatic	{011}	$(011)(0\bar{1}1)$
	Prismatic	{111}	$(111)(1\bar{1}1)$
Triclinic	Pinacoidal	All	One plane only

The indices given are for the unit cleavage forms; values other than unity are also encountered in some crystals.

*The first-order form is given; the second-order form is similar.

paper, or it can be treated with SiC particles in a Bond Wheel, such as is used in preparing single crystals for X-ray diffraction. The resulting sphere, with its bounding cleavages developed, is then examined in an optical system that is exactly analogous to an X-ray back-reflection Laue pattern, and is analyzed in the same way. Table II reproduces the Wolff and Broder data recorded in the form of stereograms. Dots show the reflections from individual crystal planes; open circles represent broader, more diffuse reflections. Solid lines indicate cleavage zones. The cleavages described in mineralogical descriptions are included above the stereograms for comparison.

Like cleavage, ***parting*** describes the presence of a planar direction perpendicular to which bonding is weak, and parallel to which cracking therefore proceeds easily. Such a plane of weakness may exist in a particular crystal as the result of events in its history. It is not subject to the laws of symmetry. Parting is also observed in polycrystalline rocks such as shales, slates, and schists. Figure 7–1 shows the crack surface of a split slate used in flooring.

Unfired ceramic specimens may also show parting parallel to some surface, not necessarily planar, along which shear took place in fabrication, such as shear from extrusion or from dry pressing, alignment at mold surfaces in slip casting, or where other forces have oriented flaky or elongated particles in a preferred orientation.

Table II.   Microcleavage Patterns

	DIAMOND	ARSENOLITE	SILICON	GREY TIN	GaSb
	C	As_4O_6	Si	$\alpha - Sn$	
	III PERFECT	III			
		SIMILAR PATTERN FOR SENARMONTITE	SIMILAR PATTERN FOR Ge		SIMILAR PATTERN FOR InAs, InSb
	Al Sb	SPHALERITE	TIEMANNITE	COLORADOITE	CUPRITE
		ZnS	HgSe	HgTe	Cu_2O
		OII PERFECT	NONE	NONE	III INTERRUPTED OOI RARE
CUBIC	SIMILAR PATTERN FOR GaAs, InP, ZnTe	SIMILAR PATTERN FOR GaP, AlP, AlAs, ZnSe, CuI.			
	FLUORITE	Au Ga_2	HALITE	PERICLASE	GALENA
	CaF_2		NaCl	MgO	PbS
	III PERFECT OOI SOMETIMES		OOI PERFECT	OOI PERFECT III IMPERFECT OII PARTING	OOI EASY & PERFECT III PARTING
		SIMILAR PATT. FOR $AuIn_2$	SIMILAR PATT. FOR LiF, KBr		SIMILAR PATT. FOR PbTe, SnSe.
	Na ClO_3	HAUERITE	PYRITE	COBALTITE	ULLMANNITE
		MnS_2	FeS_2	CoAsS	NiSbS
		OOI PERFECT	OOI, OII, III INDISTINCT	OOI PERFECT	OOI PERFECT
	$Pb(NO_3)_2$	$Ba(NO_3)_2$	MAGNETITE	SPINEL	FRANKLINITE
			Fe_3O_4	$MgAl_2O_4$	$ZnFe_2O_4$
			III, OOI, OII, I38 PARTING	III INDISTINCT "SEPARATION" PLANE	III PARTING

(continued)

Table II. (continued)

	ZIRCON	KDP	ADP	SCHEELITE	WULFENITE
	$ZrSiO_4$	KH_2PO_4	$NH_4H_2PO_4$	$CaWO_4$	$PbMoO_4$
TETRAGONAL	110 IMPERFECT 111 INDISTINCT			101 DISTINCT 112 INTERRUPTED 001 INDISTINCT	011 DISTINCT 001 } INDISTINCT 013 }
	RUTILE	CASSITERITE	ANATASE	HAUSMANNITE	CHALCOPYRITE
	TiO_2	SnO_2	TiO_2	Mn_3O_4	$CuFeS_2$
	110 DISTINCT 100 LESS SO 111 IN TRACES	100 IMPERFECT 110 INDISTINCT 111 OR 011 PARTING	001 } PERFECT 011 }	001 NEAR PERF. 112 } INDISTINCT 011 }	011 SOMETIMES DISTINCT

	BARITE	CELESTITE	ANGLESITE	$KMnO_4$	ANHYDRITE
	$BaSO_4$	$SrSO_4$	$PbSO_4$		$CaSO_4$
RHOMBIC	001 PERFECT 210 LESS PERF. 010 IMPERFECT	001 PERFECT 210 GOOD 010 POOR	001 GOOD 210 DISTINCT 010 TRACES		010 PERFECT 100 NEAR PERF. 001 GOOD
	CHRYSOLITE	FAYALITE	TEPHROITE	CHRYSOBERYL	TRIPHYLITE
	$(Mg, Fe)_2 SiO_4$	$Fe_2 SiO_4$	$Mn_2 SiO_4$	$BeAl_2 O_4$	$LiFe(PO_4)$
	010 DISTINCT 100 LESS SO	010 DISTINCT 100 LESS SO	DISTINCT IN TWO DIRECTIONS AT RIGHT ANGLES	110 DISTINCT 010 IMPERFECT 001 POOR	100 NEAR PERF. 010 IMPERFECT 011 INTERRUPTED
	SIMILAR PATTERN FOR OLIVINE				
	CERUSSITE	STIBNITE	BISMUTHINITE	MARCASITE	ENARGITE
	$PbCO_3$	Sb_2S_3	Bi_2S_3	FeS_2	Cu_3AsS_4
	110 & 021 DIST. 010 & 012 IN TRACES	010 PERFECT & EASY 100,110 IMPERF.	010 PERFECT & EASY 100,110 IMPERF.	101 DISTINCT 110 TRACES	110 PERFECT 100,010 DIST. 001 INDISTINCT

(continued)

Table II. (continued)

QUARTZ	BERLINITE	CORUNDUM	HEMATITE	Cr$_2$O$_3$
SiO$_2$	AlPO$_4$	Al$_2$O$_3$	Fe$_2$O$_3$	
10$\bar{1}$1, 10$\bar{1}$0, 01$\bar{1}$1 0001 DIFFICULT	NONE	NONE 0001 } PARTING 01$\bar{1}$2 }	NONE 0001 } PARTING 01$\bar{1}$2 }	

(QUARTZ: 10$\bar{1}$1 diagram) (BERLINITE: 10$\bar{1}$2 diagram) (CORUNDUM: 10$\bar{1}$1 diagram) (HEMATITE: 01$\bar{1}$2, 10$\bar{1}$1 diagram — SIMILAR PATTERN FOR RUBY (98% Al$_2$O$_3$, 2% Cr$_2$O$_3$)) (Cr$_2$O$_3$ diagram)

WURTZITE	GREENOCKITE	ZINCITE		TELLURIUM
ZnS	CdS	ZnO	SiC	Te
11$\bar{2}$0 EASY 0001 DIFFICULT	11$\bar{2}$2 DISTINCT 0001 IMPERFECT	10$\bar{1}$0 PERFECT 0001 PARTING	CONCHOIDAL FRACTURE	10$\bar{1}$0 PERFECT 0001 IMPERFECT

(WURTZITE diagram) (GREENOCKITE diagram — SIMILAR PATTERN FOR CdSe) (ZINCITE diagram) (SiC diagram — SIMILAR PATTERN FOR AℓI.) (TELLURIUM diagram)

HEXAGONAL

CALCITE	MAGNESITE	SIDERITE	RHODOCHROSITE	DOLOMITE
CaCO$_3$	MgCO$_3$	FeCO$_3$	MnCO$_3$	Ca Mg (CO$_3$)$_2$
10$\bar{1}$1 PERFECT 0001 } PARTING 01$\bar{1}$2 }	10$\bar{1}$1 PERFECT	10$\bar{1}$1 PERFECT	10$\bar{1}$1 PERFECT 01$\bar{1}$2 PARTING	10$\bar{1}$1 PERFECT 02$\bar{2}$1 PARTING SOMETIMES

(CALCITE: 01$\bar{1}$2, 10$\bar{1}$1 diagram) (MAGNESITE diagram) (SIDERITE diagram) (RHODOCHROSITE diagram) (DOLOMITE diagram)

WILLEMITE	PHENACITE	PROUSTITE		
Zn$_2$SiO$_4$	Be$_2$SiO$_4$	Ag$_3$AsS$_3$		
0001 MORSENET 11$\bar{2}$0 N.J. 0001 DIFF. N.J	11$\bar{2}$0 DISTINCT 10$\bar{1}$1 IMPERFECT	10$\bar{1}$1 DISTINCT		

(WILLEMITE diagram) (PHENACITE diagram) (PROUSTITE: 01$\bar{1}$2, 10$\bar{1}$1 diagram)

CROCOITE	SPODUMENE	WOLLASTONITE	ARSENOPYRITE	MANGANITE
PbCrO$_4$	LiAl (SiO$_3$)$_2$	CaSiO$_3$	FeAsS	MnO(OH)
110 DISTINCT 001,100 INDIST.	110 PERFECT	100,001 PERF. $\bar{1}$01 LESS SO	101 DISTINCT 010 TRACES	010 VERY PERF. 110,001 LESS PERFECT

MONOCLINIC

(CROCOITE: 110 diagram) (SPODUMENE: $\bar{1}$10, $\bar{1}$11, 110 diagram) (WOLLASTONITE: $\bar{1}$01, $\bar{1}$02, 001 diagram) (ARSENOPYRITE: 101, 210 diagram) (MANGANITE: 110 diagram)

From Ref. 2. Copyright by the Mineralogical Society of America.

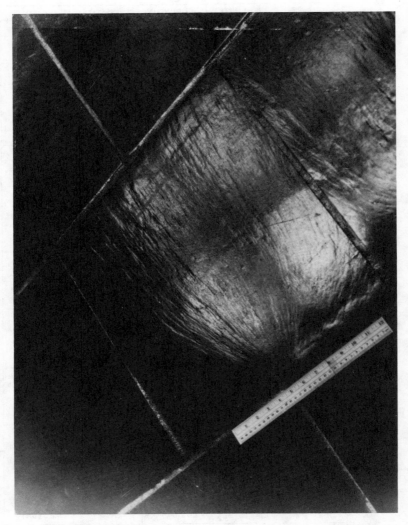

Fig. 7–1. Slate floor tile split by hammer and chisel. Note the twist hackle.

In all cases, parting can be treated as a special case of cleavage, as far as fractography is concerned.

Figure 7–2 shows the crack surface of a single crystal of periclase (MgO), which has excellent cubic and poor dodecahedral cleavages. Here cracking has followed one of the cubic cleavage planes until a twist in the axis of principal tension relative to the perpendicular to the cleavage plane initiated "stepping" in a manner analogous to twist hackle

Fig. 7–2. Central bubble and radiating cracks with twist hackle, produced by focusing the beam of a giant-pulse laser within an MgO crystal. Crossed Nicols; ×160.

formation. The breakthrough between adjacent segments, or steps, occurred along a second plane of the cleavage system. Breakthrough can occur to one side or the other, just as in isotropic media, and again barbs can appear on the fracture surface as breakthrough shifts from one side to the other. Breakthrough in both directions separates splinters, just as it does in glass.

Twist hackle in single crystals having cleavage does not necessarily imply a twisting stress field, however, but may simply indicate that the axis of principal tension is inclined to the cleavage normal. Its interpretation in terms of local crack propagation direction is the same as in isotropic materials, but the directions chosen by the running crack will be affected by the cleavage system as well as by the applied stresses.

Hackle steps may also arise when a crack running on a cleavage plane of one member

Fig. 7–3. Transmission electron micrograph of rhombohedral microcleavage surfaces in transgranular fracture of Al_2O_3. ×10,000. Courtesy of H. Kirchner and R. M. Gruver.

Fig. 7–4. Intergranular fracture surface in BeO ceramic broken at 1700°C. Dark-field illumination; × 500.

of a bicrystal meets the other member, the two having cleavage planes which are twisted with respect to one another, that is, a "twist boundary" exists between them. (A "tilt boundary" does not have this effect.)

In certain crystals, fracture—especially slow fracture—can yield a crack surface of great complexity, consisting of innumerable cleavage facets. Sapphire (alpha alumina), which shows $\{10\bar{1}1\}$ (rhombohedral) cleavage nearly at right angles (it closely resembles a cubic cleavage) is a good example. A crack running under a principal tension which is inclined equally to the cleavage planes generates a surface which, on a microscale, consists of an array of three-sided peaks and depressions, the ridges showing common alignment (Fig. 7–3). Under a principal tension which is perpendicular to one of the edges of the rhombohedron and makes equal angles with the other two, a crack may generate a striated surface. Only when a cleavage plane lies perpendicular to the principal tension is the crack surface quite flat. Stress configurations between the extremes yield crack surfaces which show features intermediate among those described.

7.2   Polycrystalline and Polyphase Materials

Crack growth shows still greater complexity in polycrystalline and polyphase materials. In addition to all of the effects met in homogeneous materials, single crystals and bicrystals are the effects of microstructural variations in elasticity, intergranular stresses,

Fig. 7–5. Transgranular cleavage in BeO ceramic showing wake hackle from intracrystalline voids and twist hackle from cleavage mismatch at grain boundaries. General trend of fracture from left to right but local directions differ in different grains. × 500.

and locally open or favored crack paths (voids, pores, preexisting microcracks, weak grain boundaries).[3,4]

At one extreme, cracks in weakly bonded or very porous materials may proceed almost entirely through voids and grain boundaries with little occasion to cross the grains themselves. This **mode of cracking** is termed **intergranular**. It results in a rough, granular-appearing surface with a correspondingly jagged profile, and the individual grains show no fracture markings (Fig. 7–4).

At the other extreme, intergranular bonding in certain materials is so strong (or the bonding phase is so tough) that a crack develops through the structure without following grain boundaries; this is termed **transgranular cracking** (Fig. 7–5). By comparison with intergranular crack surfaces it generates a smoother, more glassy surface. Yet even in this mode, the crack front does deviate somewhat at individual grains. Deviation may result from intergranular stresses which interact with the applied principal tension and alter the intensity and direction of principal stress, elastic property variations, and generation of twist hackle steps caused by cleavage mismatches at twist boundaries between grains. Thus, even transgranular fracture never develops cracks with the smoothness of crack surfaces in glass or some single crystals.

Fig. 7–6. Velocity mist hackle and twist hackle in an electrical insulator bushing tested to failure by internal pressure. Origin is at center top. Inclined illumination; ×10.

In practice, most polyphase materials crack in mixed mode, sometimes following grain boundaries, sometimes cracking through grains. Which of these is most common, in a particular material in a particular temperature range, serves to classify it as predominantly intergranular or predominantly transgranular in fracture mode. Certain materials, such as porcelain, undergo transition from intergranular to transgranular fracture when heated to a temperature at which their bonding phase begins to soften and so to become tougher.[5] In others, the bonding phase loses its integrity on heating, and the transition is to a more intergranular mode of failure. This is the case with many ceramics which fail in the creep range. Both instances fall outside the area of brittle fracture.

The fracture mode can be determined by examination of the crack-generated surface. A crack surface dominated by featureless grain outlines, or grains showing crystal growth facets or growth terraces, is identified as intergranular. Transgranular fracture, on the other hand, yields a surface dominated by fractured grains in which twist hackle from grain boundaries, and wake hackle from intergranular voids and solid inclusions, may be recognized, sometimes with primary or tertiary Wallner lines.

Examination of the crack profile can also identify the fracture mode. The crack can be seen to cross through grains or to follow the grain boundaries around them. (It is assumed that the grain boundaries are recognizable, a requirement that may be met by etching, if necessary.) This characterizes the fracture mode at the surface; whether it is the same in the interior will have to be determined by examining the crack profile on a section cut parallel to the free surface.

Following the progress of cracking in a polycrystalline material is very much more difficult than it is in glass. As much information as possible should be gained by examination with the unaided eye and at low magnifications, where, on a macro scale, surface markings can indicate the path of fracture through arrays of wake hackle and through primary Wallner lines and twist hackle. Even mist hackle may be discernible. Figure 7–6 shows the fracture-generated surface of an electrical porcelain in which the location of the origin of cracking is easily identified with the naked eye. This preliminary examination should be followed by examination of the origin region at magnifications high enough to resolve fracture markings on individual grains in order to locate the precise detail at which cracking began.

The optical microscope is frequently best if the grains are not too fine, particularly when used with plastic replicas (see the later section on Techniques). The great depth of focus of the scanning electron microscope makes it useful for dealing with surface markings. Unfortunately, this depth of focus can lead to confusion when a complex of crack planes is involved. The problem is usually acute only at some critical site and it can be resolved by photography of stereoscopic pairs, a simple and not very time-consuming procedure which not only provides three-dimensional information but also adds enormously to the quality of the perceived image. It must be admitted that certain details, such as broad primary Wallner lines, precavitation hackle, and scarps, may not be discernible with the scanning electron microscope, even though they are readily observable with the optical microscope under appropriate illumination.

8. Procedures and Techniques

8.1 Records and Check Sheets

The failure analysis of brittle materials should be systematic, lest inadvertent omission of some step should later be embarrassing. Much of what is to follow under this heading will surely be obvious, but the obvious is not always remembered at the crucial moment. Recall that able and experienced aircraft pilots are required, prior to take-off, to go through printed checklists. In the absence of a formalized system, it may possibly be that even we who routinely analyze material failures could overlook a critical step in procedure.

As with many professional activities, it is best to institute the most formal procedure, from which it is easy to relax procedures to fit the less-demanding situations.

It is well to anticipate the uses to which records may ultimately come to be put. Wild surmises have no place in such documents and are better left unrecorded until confirmed, or at least identified as simply "ideas to be considered." This will forestall the possibility that the notation may later be thrown up to the analyst as evidence that he or she "changed his or her mind" for some sinister reason.

In legal matters, the analyst may be required to certify that all parts of the analysis were conducted personally by him or her or by someone "under his or her direction and control." It is a good idea to have the records indicate this information, if only by a preface stating that all analyses were performed by the author except where explicitly recorded otherwise.

8.2 Acquisition Records

It is essential that the analyst record full information about the specimen he or she is to analyze. An acquisition sheet is helpful, drawn up to suit the particular activity. No one form will be appropriate to all assignments, but certain basic requirements must be met. These include:

1. From whom the specimen was received with date and time of receipt.
2. Identification of the specimen in terms of the incident in which it failed.
3. Number and total weight of fragments.
4. Name of person requesting the analysis, and where he or she can be reached for clarification, if necessary.
5. Purpose of the analysis, that is, what information is requested.
6. Restrictions on treatment of the specimen during analysis.
7. Type of report expected and to whom it is to be transmitted.
8. Deadline for report.
9. Disposition of specimen after analysis.
10. Reference to pertinent documents.

Such a sheet should accompany the specimen at all times, although a duplicate copy for filing may be desirable. Assignment of an analysis number may be useful, but it should not be used in such a way as to clutter the files of those requesting analyses.

8.3   Examination and Analysis Records

Similarly, a check sheet may be essential to efficient analysis, where specimens of a common type are routinely handled. Such a sheet lists the observations to be made, with space for the corresponding data. The following will simply indicate some of the possible items for routine entry.

1. Quality of material.
2. Quality of fabrication.
3. Quality of processing.
4. Conformity to dimensions.
5. Condition of specimen, including wear and damage from abuse.
6. Fractographic analysis.
7. Photographs taken, with their negative file numbers.

Others may be added, and any of these may be subdivided if convenient to ensure that no essential detail will be overlooked.

8.4   Graphic Records

Graphic recording of analytical results is generally valuable, not only for use in reporting to others but also for refreshing the analyst's mind in future discussions, sometimes years after the analysis. Photographs, television recordings, sketches, replicas, and even rubbings can be effective for these purposes.

The choice of a photographic system is a matter of relative priorities among convenience, reliability, cost, and quality. Large-format view cameras can provide the ultimate in photographic quality. Instant-development film affords the great comfort of a photo in hand as a guarantee against photographic errors, malprocessing, or loss of film in handling. Small-format cameras, especially 35 mm cameras with "macro" lenses, are the least expensive and the most convenient, especially where modern same-day commercial processing is available. But there is an element of risk with them which may be decisive in those situations where retakes will be impossible.

Whatever photographic system is chosen, the advantages of stereoscopic pairs should be considered. The technique is extremely simple and quick. After the first view is snapped, the camera is simply moved a short distance to one side, sufficient to swing it through an angle of six to ten degrees, and a second photo is taken. Together they constitute a "stereopair," which can be viewed in a stereoscope such as those available commercially for surveying, geology, or radiography. A satisfactory stereoscope can be constructed easily from four mirrors, arranged so that each eye looks into a mirror that is laterally inclined 45 degrees; the line of sight then meets a second mirror which redirects it in the original direction toward one of the stereopair photographs.

Stereopairs not only provide three-dimensional information, but an image which is vastly superior to either of its members. (Evidently the brain has an image-processing function by which it routinely constructs from two imperfect pictures a composite image that is formed from the best features of both.)

Television recording is becoming more and more useful, as equipment becomes better and cheaper. Its greatest value at present lies in the ease with which it can display observations to a group, making use of movement to convey three-dimensional impressions and to relate the several aspects of a specimen to one another.

For both photography and television recording, proper lighting is critical for obtaining good results. Perhaps the best results can be obtained with well-controlled artificial lighting, but it is hard to beat natural lighting, close to a window, with light-colored walls and ceiling to provide the fill. It is well to think out the impression that is to be communicated, so that significant details can be identified by placing numbers, arrows, and so on at appropriate places on the subject. A label in the picture, recording the subject, date, and possibly a serial number, is useful in keeping track of records.

Sketches have some advantages over photographs as quick memoranda and as a means of communicating observations which are difficult to photograph or are cluttered with irrelevant details. They are especially appropriate for showing fragment shapes and surface details simultaneously with the significant details of the crack surfaces which bound them.

Sketches are useful too for identifying fragments by number and position in the reassembled specimen. They fill a particular need in describing a specimen such as a bottle, whose entire surface cannot be seen in one view. Tracing paper can be wrapped around such a shape for sketching (it is well to include some overlap) and then spread flat for the addition of notes, such as thicknesses and indications of crack development. Like all records, sketches should be properly labeled and dated.

8.5   Cleaning the Fragments

Specimens are often received in a soiled condition which interferes with inspection and fractographic examination and makes it desirable that they be cleaned. Before cleaning, it should be considered whether the soil itself may have some bearing on the analysis. Examples of this: the crack origin surface which is soiled with a substance which could have got there only prior to the failure, demonstrating that it represented an earlier event; the oil lamp fragments with residues indicating use of an improper fuel; the porcelain insulator fragments with coatings of white lead showing the slow action of a leakage current over a long period prior to final failure. If any such circumstance is suspected, it is wise to photograph the evidence and to preserve samples of the soils or residues for analysis, should that be found appropriate later.

If a solvent must be used for collection of a sample of soil, it should be as pure as possible. A specimen of the solvent alone should be stored separately. Both should be in containers which will not contaminate them.

Particulate materials can sometimes be removed by gentle brushing, or by careful scraping with a plastic tool. In more difficult cases they can be removed and preserved by pressing cellulose tape, softened by ten to fifteen seconds immersion in acetone, against the fragment surface. After drying for a few minutes until crisp, the tape is stripped off and stored.

8.6   Fractographic Analysis

The fractographic analysis itself usually requires that the primary origin of cracking be located and studied. It is not enough to locate simply a point from which cracking extends in all directions; it is common to have several of these. It is essential to identify the origin where cracking first began (in special cases it may have begun at several points simultaneously and independently). To this end it is necessary to trace the development of cracks from every origin until each one vents to a free surface, aborts, or terminates in a crack belonging to its own or to an earlier system.

It is well to begin by inspecting each fragment edge before attempting to reassemble. In the first place, this is a great help in reassembly, in the same way that the picture on

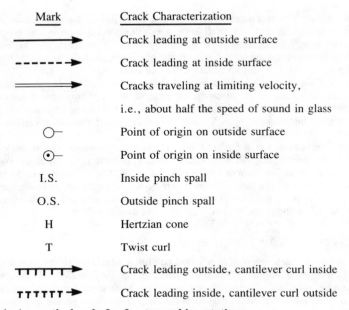

Fig. 8–1. A practical code for fractographic notations.

a jigsaw puzzle is a major help in putting it back together. Second, the information gained on each edge can be recorded by marking the free surface so that it is available after assembly to interpret the history of crack development.

Notations on the free surface can range all the way from a simple arrow indicating the direction in which the crack traversed an edge, to elaborate codes indicating the surface at which the crack was leading, types and prominence of Wallner lines, propagation at terminal velocity, evidence of cantilever stress or shear stress, and the presence of spalling. As an example, Fig. 8–1 shows a code which has been found to be practical. Such notations can be made with a toluene-based or alcohol-based ink (preferred) or a wax pencil, both of which can easily be removed later if desired, or by India ink penned on the surface of cellulose pressure tape which can be peeled off or soaked loose with water. Since your code may well evolve as experience with your peculiar problems is gained, it is well to keep track of what code was used with a particular specimen.

Additional notations can be made, preferably in other colors, tracing the continuity of crack systems in order to determine the sequence in which they occurred and thus to establish which was the primary origin.

Fracture origins should receive particular attention during this examination prior to reassembly. With each, the character of the origin flaw should be noted, the crack front shape during early development, the flatness or curvature of the crack surface around the origin, the shape and "radius" of the mirror region, the ware thickness, and any connection between the origin flaw and other surface damage in the vicinity.

In reassembly, the pieces can be attached to one another with cellulose pressure tape, one corner of each strip being tucked under to aid in removal. The clear tape is preferred over the matte, since it is less conspicuous in a photograph. If the specimen is thick-walled it will easily support its own weight during manipulations; otherwise it may be necessary to use an appropriate support to prevent sagging or collapse. Application of tape to both sides of the fragments will contribute to the stability of the assembly.

8.7   Optical Examination

In general, optical examination should be begun at low magnifications and increased only as necessary to meet the needs of the analysis. The sequence begins with the unaided eye. In some instances, such as in studying fractures in rocks and other massive objects, it will pay to stand well back, lest the major features be overlooked in preoccupation with the minor ones.

The hand lens, held close to the eye, is useful in the magnification range from × 3 to × 10. Next is the stereoscopic microscope with magnifications as high as × 200 (zoom optics are especially convenient). The three-dimensional capability of this instrument makes it comfortable to use for understanding the spatial relationships of the fracture features, but some of the time, as when examining glass fractures, the stereoscopic microscope must be used one-eyed, to avoid conflicts in illumination between the two images.

Illumination is a critical matter. There are two principal modes of illuminating a specimen, each fitting a particular type of material. With granular materials, fracture markings are most clearly visible when they are illuminated at a glancing angle, just as objects in the landscape are brought into sharp relief when the sun is low in the sky. With these materials the stereoscopic microscope should be used with both eyes open, although it is prudent to rotate the specimen in all directions with respect to the light, in order to guard against spurious impressions.

With glass, however, subtle markings are brought out best by lighting the surface at the edge of the reflecting condition. The featureless area, or background, can be tilted so that it does not quite reflect the light source, and so it appears dark. Surface perturbations then reflect the light and appear bright against a dark background. This is called near-dark-field illumination; it is the most sensitive condition. Alternatively, the background can be set to reflect just the light source and therefore appear bright; the surface markings do not reflect and so appear dark. This is the so-called bright-field condition. With the stereoscopic microscope, it happens from time to time that one eye sees the specimen in bright field, the other in dark field. The combined images are then unintelligible and one must move one eye aside to observe effectively.

At low magnifications, it is convenient to hand-hold the specimen in the stereoscopic microscope. At higher magnifications, the specimen must be supported more steadily. A lump of modeling clay on a glass slide holds small specimens conveniently and allows easy manipulation. It does, however, soil the side of the fragment which is embedded in it. Cleaning afterward is easy, but if this is unacceptable, the specimen may be placed in a clamp so that only the clamp is in contact with the modeling clay. Some such device is essential when linear measurements (of origin flaws, for example) are to be made, as with an eyepiece micrometer.

A more elaborate device, for manipulation of fragile specimens, consists of a hollow hemisphere supported in a ring, the specimen being located at the sphere center (Fig. 8–2). This provides all degrees of freedom of tilt and rotation without disturbing the specimen itself.

For specimens requiring higher magnification or greater contrast than are attainable

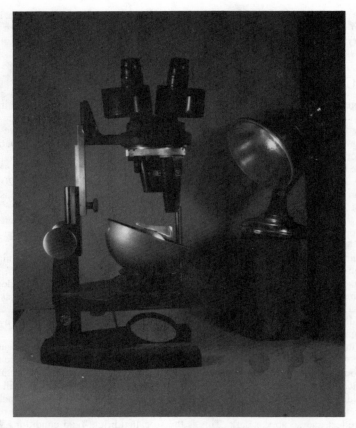

Fig. 8–2. A rotating specimen mount consisting of a hollow hemisphere resting on a ring.

with the stereoscopic microscope, optical microscopes are available with useful magnifications to about × 1000. Differential interference contrast optics (these are familiarly known by the name Nomarski, whose particular version may not be the most satisfactory, however) are to be highly recommended. They provide excellent contrast without sacrifice in image sharpness, and are available for both transmitted and reflected light systems.

With glass specimens, good results can be obtained in both reflected and transmitted light. High-quality photography in reflected light may require that the crack surface be silvered to eliminate internal reflections, but such reflections are not a problem in visual examination or in routine photography.

Opaque or granular materials can be observed by incident or reflected light. Dark-colored or opaque materials can be examined successfully in this way, although there is danger of scratching expensive microscope objective lenses. White materials, on the other hand, may be nearly impossible to work with in reflected light. Less than 10% of the incident light (slightly more from certain ceramics) is reflected from their surfaces to form a coherent image; the remainder is transmitted through the surface, is scattered at sub-surface grain boundaries, and much of it returns to fog the image and obscure its details. Such fracture surfaces can be examined at high magnification through the use of plastic replicas, whose preparation is to be discussed in a later section. Transmitted light is preferred for studying replicas because it is more readily controlled.

8.8   Replication

The use of replicas to replace direct examination of fracture surfaces follows the lead of the transmission electron microscopists who adopted replication to circumvent the problem of the low penetrating power of the electron beam. Optical microscopy hasn't this problem, but nevertheless finds a number of important advantages in replicating:

1. The replica is an easily stored record of the critical part of a specimen which is too large to store or must be otherwise disposed of, as in the case of legal evidence or in circumstances where the specimen is to be tested destructively.

2. Optical examination is facilitated by the thinness of the replica and by the absence of disturbing subsurface details. Either transmission or metallographic (reflection) optics may be used with it. The soft, pliant plastic is far less dangerous to the objective lenses than the original specimen if accidental contact is made.

3. Negative details such as pores and cracks in the specimen are converted to positive details in the replica. In this form they are more effectively imaged in the scanning electron microscope, which sees into cavities poorly.

4. Curved surfaces are made more accessible to high-magnification study, since they can be flattened in mounting the replica.

For optical use, the replication processes are very much less demanding than those practiced in electron microscopy, since transparency to an electron beam is not required. Ideally, replication should be simple and fast, and it should yield a replica that is transparent, pliant, strong, and permanent. The choice of a replication method depends on the specimen, the facilities at the site, and the urgency of the matter. Better methods will certainly appear, but for the present three are available to cover practically every need. Each has its own advantages and shortcomings.

Poly(vinylchloride) (PVC), a thermoplastic commonly available in plastics supply houses in the requisite unpigmented sheet form, is melted in contact with the specimen, cooled, and stripped to give a strong, pliant replica. The temperature of heating must be determined by experiment, since it varies with the formulation of the sheet. The precise procedure must be adapted to the size and nature of the specimen. Typically, it consists of heating the specimen to a temperature of about 180°C on a hot plate or in an oven; a strip of the sheet is then laid on its surface and quickly rolled with a smooth Teflon* rod to work out entrapped air. After cooling until set but not cold, the replica can be stripped off the specimen by a quick tug. After trimming away the unwanted rims, the replica can be mounted with tape on a microscope slide or on a projector slide cover glass. As with all replicas, it is best mounted impression-side down, so as to protect the impression from dust and so that it appears from above as a positive. This also makes it possible to reach the relevant portions with high-magnification microscope objectives without interference from the edge folds, and to use oil-immersion objectives if necessary.

Small specimens may have too low a heat capacity to be worked in this way. They

---

*E.I. du Pont de Nemours & Co., Inc., Wilmington, Del.

can be replicated by clamping them to a ring stand or other movable support so that they can be lowered quickly to a predetermined level into a pool of molten PVC, formed by brief heating on an oversized microscope cover slip on a hot plate. This will then be inverted so that the impression surface is seen through the cover slip. The unpleasant nature of the fumes from hot PVC make it necessary to work in a fume hood.

The most satisfactory replicas of all are made from an unpigmented two-part silicone elastomer, such as Sylgard 182[†] from Dow-Corning. The monomer is thoroughly mixed with 10% curing agent in a disposable 8-ounce plastic container and is deaired by a few cycles of evacuation to 28 inches of mercury to remove entrapped air bubbles. The specimen is prewarmed to about 100°C and is clamped on a movable stand to place the critical surface within 1 mm of a large microscope cover glass. The two-part mixture is then placed on a hot plate at about 100°C and the specimen lowered quickly into it, care being taken to avoid trapping air. After about 10 minutes, the assembly can be removed, cooled, and stripped by teasing up the edge of the replica with a blunt knife. It is examined by focusing through the cover slip, to avoid interferences of the meniscus with the objective and to permit the use of oil-immersion objectives.

These silicone "rubber" replicas are nearly ideal, but their mechanical weakness can cause trouble in stripping them away from the specimen, especially if its surface is rough. Additionally, the requirement that the specimen be heated, if curing is to take place in a reasonable time, may not always be acceptable.

If replication is to be done at room temperature, cellulose acetate tape, such as is used for transmission electron microscopy, may be used. After 10 to 15 seconds immersion in acetone, the tape is laid on the acetone-flooded surface and pressed firmly against it for two to three minutes using a smooth, featureless pad such as cast silicone rubber or a PVC sheet backed with folded cloth. The pad can then be gently separated and drying allowed to continue for another ten minutes before it is stripped off with a quick pull. The resulting replicas tend to be stiff and awkward to mount but they give reasonably satisfactory results.

In all work with replicas, it is necessary to be alert for artifacts, such as deformed bubbles or trimming marks from knife or scissors.

8.9   Electron Microscopy

The resolving power of the optical microscope is limited to about half the wavelength of visible light, that is, to about 0.2 μm under ideal conditions. The very short effective wavelength of high-energy electron beams, on the other hand, gives them virtually unlimited resolving power and so brings the tiniest constituent grains of a material within the range of observation. Consequently, the electron microscopes and their related analytical modifications are indispensable in dealing with many of the problems associated with modern materials.

The scanning electron microscope is the most versatile and least expensive of these. It may or may not be helpful in a particular failure analysis. Its greater maximum magnification and its great depth of focus make it attractive in comparison with optical instruments, especially for photography. It permits excellent over-the-edge photographs, in which the fracture surface appears at the same time as the adjoining free surface. Yet it is time-consuming and expensive, and the modest size of the specimen chamber precludes examination of large specimens, except through the use of replicas. Also, certain

---

[†]Dow-Corning Corp., Midland, Mich.

fractographic details, such as precavitation hackle, scarps, and some Wallner lines, may not be discernible with the electron microscopes, even when they are readily seen with the optical microscope under appropriate illumination.

The transmission electron microscope is rarely used in failure analysis except in research, where its capacity to locate and identify extremely small particles, and to describe dislocation arrays and stacking faults, has contributed a great deal to the understanding of materials and the fundamentals of their mechanical properties. It will not be discussed further here, but there is an enormous literature on the subject, including the proceedings of periodic international conferences.[1]

8.10   Evaluation

At some point the analyst may have to face the dilemma that his or her analysis is at variance with the report of an eyewitness to the failure. That should surprise no one.

The general perception that the account of an honest eyewitness is the best of all evidence is quite mistaken. The fact is that, especially when an event is sudden and traumatic, the eyewitness is wholly unreliable. This is doubly unfortunate. First, many an event is falsely reconstructed because of statements from mistaken eyewitnesses. Second, many an eyewitness has been thought to be lying, because of unwitting errors in his or her testimony.

This is not the place to discuss why this is so. Apparently it has something to do with our tendency to have little or no memory of the dramatic moment itself but rather to later reconstruct in our minds what must have happened. It is apparently a short step from the conclusion that something must have happened to the absolute conviction that it is what we ourselves saw. That is especially true of events which we recount and go over in our minds again and again.

The witness testifies, for example, that the bottle exploded and then was dropped, when it is quite clear from fractographic analysis of the fragments that the bottle was dropped first.

The problem of the unreliability of the eyewitness is by no means restricted to legal situations. The industrial workman or device operator may be completely oblivious to errors in his or her operation. The fractographic report should be phrased to enable cure rather than to establish blame.

The point here is to advise the failure analyst to virtually ignore the testimony of eyewitnesses in the process of reaching his or her conclusions. And if he or she finds from the physical evidence that the eyewitness is mistaken, he or she should not be tempted to impugn the competence or character of the witness but merely to reflect on the innate fallibility of the human mind. Neither should the analyst allow himself or herself to be shaken by the reports of eyewitnesses, no matter how positive. The physical evidence traced by the running crack is the infallible and incorruptible witness!

8.11   Reporting

Communication of the results of failure analysis is a demanding exercise in technical writing, and it must be taken seriously in proportion to the anticipated impact of the results. Clear, explicit statements of procedures and findings are called for, with adequate arguments to support the reasoning behind the conclusions reached.

Report format will be dictated by official policy and personal taste, but it should be arranged to accommodate essentially the same points covered in the section on Records above.

Drawings and photographs should be identified, not simply by number but with sufficient information in each caption to enable the reader to understand the main significance of the figure without going to the text.

The section of the report entitled Conclusions or Summary demands special attention. It cannot include everything, of course, but the statements made there should stand by themselves as far as possible. The reader should not be required to go back to the pages of the text to find out what is being said, although he or she may wish to go back for additional details.

9.0 Common Conditions of Failure

The basic principles underlying the failure analysis of brittle materials have been reviewed to the extent that presently seems possible. The present section will indicate how they can be expected to appear in certain typical failure conditions in order to underscore their relevance and to emphasize that they complement and reinforce one another.

In all cases it will be necessary to refer to the nature of the stresses responsible for failure. But it should be recognized that stresses at the running crack front are not those of the original stress configuration, since they are continuously subject to changes brought about by the presence of the crack system generated up to that point.

9.1 The Center-Heated Plate

A common form of failure results from plates or slabs heated in the center while the edges remain cool. A classic example is the large plate-glass (or float-glass) window heated by the sun's radiation, the edges being protected from the sun's rays by the framing, or being kept cool by abutting masonry. Center heat expands the central portion, and the nonexpanding edges are thereby subjected to in-plane tensile stress parallel to the edge. (The edge is precisely the location in which stress-raising flaws are most likely to be present, introduced either by imperfect scoring and breaking or by careless handling prior to framing.) Under these circumstances cracking begins at the edge or very close to it, running perpendicular both to the edge and to the face.

The crack runs flat and straight at first and then may curve away. Very often it develops a meandering path (see Section 3.2). If the stress in the glass at failure is low, that is, below about 0.22 Pa (1500 psi) according to Orr,[1] velocity forking is not to be expected. At higher stresses forking may occur, but only within a short distance of the edge, within two inches in the case of large windows. Because of the uniform tensile stress field the mirror region is circular about the point of origin. By contrast, out-of-plane stresses caused by wind pressure, by a mechanical bump, or by twisting or pinching in the frame lead to a noncircular mirror, and the crack develops a surface which is not perpendicular to the face. Figure 9–1 summarizes some of the characteristics which distinguish between failures of thermal and mechanical origin. Figure 9–2 indicates where the mirror radius is to be measured and the relationship between mirror radius and stress in the glass at failure.

When the center-heated plate is not flat but domed, the expansion of the heated center can be relieved not only by pressing outward against the edges but also by bulging farther from the plane of the edges, that is, by decreasing its radius of curvature. This bulging develops a tensile stress in the convex surface and may be sufficient to initiate cracking there. The crack advances most rapidly at the outside surface and is retarded in its development through the center at first, but to a lesser degree as it approaches the edge.

All of these considerations apply equally to a plate that is initially warm and is cooled, the cooling taking place more rapidly at the edges. Such cases are commonly encountered in ware in the cooling cycle in ceramic kilns. The exposed edges lose heat

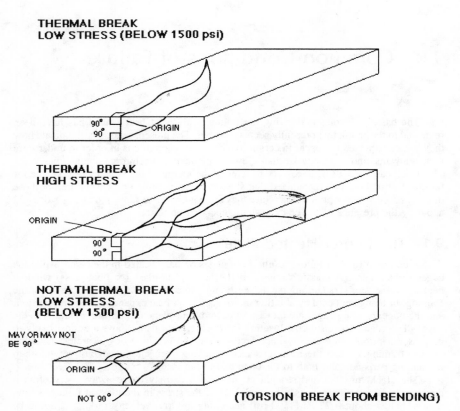

THERMAL BREAK
LOW STRESS (BELOW 1500 psi)

90°
90°
ORIGIN

THERMAL BREAK
HIGH STRESS

ORIGIN

90°
90°

NOT A THERMAL BREAK
LOW STRESS
(BELOW 1500 psi)

MAY OR MAY NOT
BE 90°

ORIGIN

NOT 90° (TORSION BREAK FROM BENDING)

Fig. 9–1. Recognition of thermal (in-plane) and bending (out-of-plane) breaks in window glass. Courtesy of PPG Industries.

more quickly than the interior, both by radiation and by loss of heat to the circulating atmosphere; accordingly, cracking may originate in the edges. The effect is most likely to occur in the creep range and especially in temperature ranges where thermal shrinkage is unusually large, as in the neighborhood of the quartz inversion in whitewares.

9.2 The Pressure Vessel

A cylindrical pressure vessel experiences two principal tensile stresses at any point in its wall, an axial stress parallel to the cylinder axis, and a hoop stress in the circumferential direction. For thin-walled vessels, the hoop stress σ_H is taken to be simply:

$$\sigma_H = D{\cdot}P/2t$$

where D is the internal diameter, P the pressure, and t the wall thickness. The axial stress is exactly one-half of this, and so failure occurs under hoop stress and develops cracking parallel to the cylinder axis.

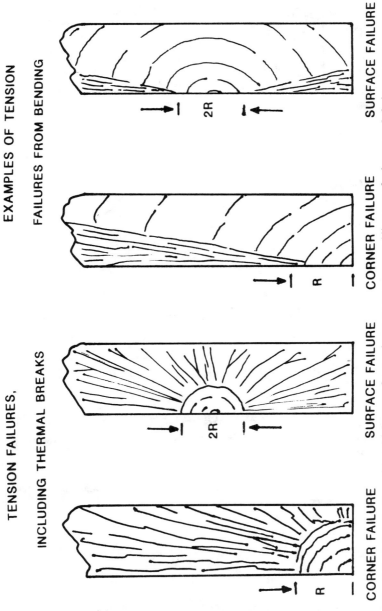

Fig. 9–2. Determination of breaking stress from mirror measurements: (A) types of mirrors and their measurement. Courtesy of PPG Industries.

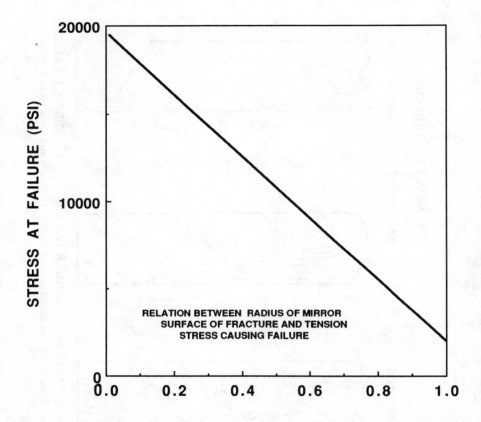

RADIUS OF MIRROR AT FRACTURE (INCHES)

Fig. 9–2 (*B*) **Relation between mirror radius and stress of failure. Courtesy of PPG Industries.**

Thick-walled pressure vessels develop a hoop stress at the inside surface that is substantially higher than at the outside:

$$\sigma_H \text{ inside } = p(a^2 + b^2)/(b^2 - a^2)$$

$$\sigma_H \text{ outside } = 2p \cdot a^2 /(b^2 - a^2)$$

$$\sigma_H \text{ inside}/\sigma_H \text{ outside } = (a^2 + b^2)/2a^2$$

Carbonated beverage bottles are good examples of the sequence of crack development in thin-walled pressure vessels under bursting pressure (Fig. 9–3). Although the

Fig. 9–3. Bottles burst by internal stress at successively higher pressures left to right.

hoop stress is slightly higher on the inside surface, the origin of cracking is almost always outside. There the most serious flaws are to be found, both because of the nature of the manufacturing process and because the external surface is exposed to damaging contacts, whereas the inside is protected.

Since the initial stress condition is one of practically uniform pure tension, the crack front can be expected to expand through the thickness as a semicircle and to develop a flat surface perpendicular to the wall. In practice, this will be influenced by the shape of the origin flaw. If this is long in the direction of the cylinder axis, the crack front will spread initially as an ellipse. If the flaw is hooked or otherwise blunted at one end, that end may not contribute to the development of cracking; failure will begin at the other, sharp end and sweep from there around and past the inactive end.

As the cracks run, they begin to lead at the inside surface, as faint Wallner lines and the first appearance of mist hackle indicate. This is not completely unexpected, since the tensile stress is a few percent higher at the inside. The effect is very much accentuated, however, by the initial presence of stored compressive stress at the outside and balancing tension at the inside surfaces, the result of the rather rapid cooling in the annealing process. (This prestressed condition is by no means undesirable, since it requires that the surface compression must first be overcome before net tension can appear at the flaw sites. But it must be kept within bounds if undesirable stress concentrations at the heel, the transition region between the cylinder wall and the back, are to be avoided.)

The first velocity forking takes place with a maximum subtended angle of 90° between the outermost branches. The number of forks is linearly proportional to the stress in the glass at failure, or to the bursting pressure, other things being equal, a useful relationship if calibrated for the size and design of the container in question. (See Section 5.3.)

During cracking by internal pressure, the crack surfaces are characteristically simple. Primary Wallner lines may arise from surface flaws or design features. Secondary Wallner lines accompany the mist hackle prior to forking. Tertiary Wallner lines from the initial snap are limited to the near neighborhood of the origin. Twist hackle is generally absent. The crack surfaces form at right angles to the free surface, until near their terminations, where the cracks may hook, become inclined to the perpendicular, and develop center-line shear hackle. The last stages of breakup from high internal pressure are various and unpredictable.

The outward-acting force from the internal pressure exerts a bending moment on the fragments and may cause cross-cracking of the fragments in a circumferential sense, the corresponding crack fronts leading strongly at the inside surface and developing cantilever curl at the outside. These circumferential cracks usually originate from the edges of the primary, radiating cracks, especially in mist hackle regions. Occasionally they originate independently at the inside surface. In every case they terminate in members of the radiating system, indicating that they are secondary effects.

Circumferential cracks may include members which link the outermost members of the upward- and downward-running systems in such a way as to yield fragments resembling the wings of a butterfly. They do not always form in pressure bursts, and since they can be produced in other failures also, their appearances must not be taken as decisive evidence of any particular mode of failure.

Failure of cylinders with flat end faces may begin at the outside surface of the end face, where tension is active in all directions. The cracks vent to the sides where they may proceed directly up the sidewalls or may curve to follow the heel for some distance before running axially with exactly the same characteristics as noted above for crack systems originating in the sidewall. Forking may occur at all stages.

Bending stress at the heel may lead to cracking from an inside origin there, particularly if the transition from sidewall to base is sharp, that is, the inside radius of the curve of transition is small, or if unusually effective flaws are present in the inside heel. The initial crack runs perpendicular to the cylinder axis.

Finally, failure at a domed surface, such as that approximated at the shoulder of some

Fig. 9–4. Impact damage from a blow of moderate intensity showing radial and circumferential damage.

containers, takes place under a tensile stress in all directions, so that the initial crack can proceed in any direction prior to exiting from the domed region. Stress in it is equal to the axial stress in the cylindrical region and so is just one-half of the hoop stress in the cylinder wall; bursting from an origin in the shoulder is therefore seldom observed.

9.3   Impact Failure of a Sheet or Slab

Impact, in which the momentum of a moving body is abruptly converted into a contact force, provides one of the most common and the most dangerous conditions encountered by brittle materials. Earlier it was noted that impacts which are insufficient to cause failure promptly can inflict damage which leaves the specimen susceptible to failure under subsequent loading.

The Hertzian cone popped out of the back of a window pane by high-velocity impact from a hard, rounded object against its face may or may not be considered to constitute a failure. To be sure, the window may remain in place substantially intact, but for reasons of appearance or because of air leakage, the hole may be unacceptable. Increased velocity of impact leads to development of cracks radiating outward from the impact center (Fig. 9–4), leading at the back surface. At still higher velocity the slender fragments separated by these radial cracks are pressed inward at their tips, so that they crack across to form circumferential cracks which lead at the front face and which show cantilever curl toward the back face. In the meantime, intrusion of the projectile into the cavity opened by the cone leads to lateral forces which pinch off flakes of the specimen at both front and back surfaces, generating the spall characteristic of this stage of impact damage.

Impacts by sharp or jagged projectiles result in the same type of damage, with the exception that the Hertzian cone may be absent or distorted. The initial stages were discussed in Section 4.2.

Heavy blunt impacts do not raise contact stresses sufficiently to initiate failure at the circle of contact on the front face but, by deforming the sheet inward, raise tensile stresses at the back face, in directions which are determined by the buckling tendency of the sheet. Failure may then begin from a point in the area opposite the center of contact, if a suitable flaw is present. However, tensile stresses appear all along the bend ridge which may form in a direction established by the shape of the sheet, the arrangement of its supports, and the location of the center of impact. Not infrequently, failure may begin at a flaw in the edge, run inward to the center of impact, and there trigger breakup in the form of a pattern of radiating cracks, leading on the back surface. The fragments thus separated may be broken off by bending stresses exerted on their inner corners by the still-active impact force; the resulting circumferential cracks lead at the front surface and show cantilever curl toward the back. Their origins are often at the edges of the radial cracks or they may originate independently, from surface flaws. Most of them terminate at both ends in the first radial cracks that they encounter, but because they lead at the outside surface while the radial cracks lead at the inside, there are instances where members of the two families intersect one another and continue beyond.

A characteristic of the radial crack system is the tendency to decelerate with distance from the center of impact. Velocity forking in impact failures usually occurs early or not at all. The resemblance of the pattern to a spider web is striking. All of the characteristics of heavy blunt impact are shared by failure from simple pressure against a limited area of the sheet.

9.4   Impact Against a Cylinder or Dome

Impact against a cylinder, such as the sidewall of a bottle, has much in common with impact against a sheet, with modifications resulting from the differences in compliance of the two shapes. The force of impact has the effect of deforming the cylinder inward at the center of impact and of generating a corresponding outward bulging in the regions roughly 45° around the cylinder on either side of it.[2] Thus large areas of the outer surface are brought into tension while the inner surface is brought into corresponding compression. While the tensile stresses in these areas are lower than at the contact circle in the case of impact against a hard material, the large areas subject to such tension make the likelihood of encountering a potential origin flaw more likely. Similarly, the tension at the inside surface opposite the center of impact is also larger than in the bulged-out areas but, in the case of bottles, flaws are unlikely there, and cracking is likely to originate at one side or the other or, sometimes, both. Such origins are called *hinge origins* and the systems of cracks extending from them are called *hinge breaks* or *hinge systems* (Fig. 9–5). Their members lead on the outside surface and show cantilever curl at the inside. A crack from a hinge origin may run to the impact center and there trigger a typical impact "spider web"; such a crack is termed a *leader crack*.[3] Hinge origins are under tension from all sides and so may run in any direction, often forking repeatedly in both directions to generate sprays of cracks. Even when not including a leader crack, they may form simultaneously with the central impact system. Members of the hinge systems may, in their early development, act to terminate the outermost radial members of the central

Fig. 9–5. Hinge origin (center) in a bottle broken by impact against the surface at the right.

system while the innermost, that is, the most nearly axial, of the members of the central system run faster and terminate the central members of the hinge systems.

Because of the curvature of the cylinder, the region of the impact center is typically spalled, to an extent not common in any but thick sheets or slabs. The fragments, collapsing inward, are forced against one another by reason of the curvature, pressing flakes from each others' surfaces, both inside and outside. The resulting pattern is strikingly symmetrical around the cylinder axis, the extent of spalling being greatest in the lateral direction while the density of radial cracks is greatest in the axial direction. Subject to calibration, the frequency of cracking in the axial zone can be used to estimate the severity of impact, but because of the many variables, including the shape and hardness of the impacting body, the usefulness of this relationship remains to be demonstrated. (See also Section 5.3.)

Fig. 9–6. Tumbler with symmetrical radiants from a Hertzian conoid.

Failure by impact against a domed surface, such as approximated by the shoulder of some bottles, can be seen as an extension of the principles just stated. Because of the intrinsic rigidity of the dome, cracking is unlikely to occur under any but high-velocity, hard-surfaced impacts, and origins are to be expected at the contact circle on the outside only. Hinge origins are not to be expected, in general, although the possibility cannot be ruled out, and each case must be separately evaluated. Heavy spalling can be anticipated, in any case.

Impact against the outside of a vessel's heel often develops a classic Hertzian conoid with radiants of striking symmetry (Fig. 9–6).

9.5   Waterhammer I. The Arrested Liquid Column

When liquid in motion is suddenly arrested, it presses against the walls of the containing system and deforms them, while the kinetic energy of motion becomes converted into elastic energy of the container. This is one kind of water hammer. Internal pressure in the liquid rises and the container walls are correspondingly stressed, the maximum being reached at the instant that the liquid velocity is reduced to zero. Pressures resulting from this effect can be large. If a one-quart beverage bottle is dropped only one foot, the pressure within it is increased by about 0.05 Pa (360 psi) when it strikes a stiff surface. The maximum pressure is felt close to the base. Half-way up the container the pressure rise is only half as much. These pressures are of very short duration and the

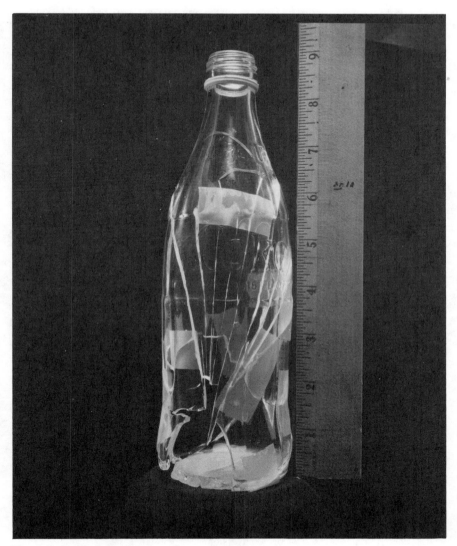

Fig. 9–7. Bottle burst by dropping on its base.

ability of the material to withstand the stresses is much greater than if the loading were prolonged.

In terms of failure analysis, this means that cracking is most likely to begin close to the heel of a container dropped on its base (this will be all the more likely because of mechanical forces occasioned by impact with the arresting surface). Velocity forking can be expected to occur with decreasing frequency as cracks extend farther and farther upward from the base (Fig. 9–7).

9.6 Waterhammer II. The Collapsing Void

Water hammer of a second kind can occur when liquid partly fills a container, the free space being at very low pressure. A sudden surge can cause the liquid to cross the container and, without the damping effect of a gas cushion, to slap against the wall with a hammerlike effect. The resulting force is unexpectedly great, considering that the linear motion of the liquid is on the average insufficient to exert much pressure on the wall. But amplification takes place, because the first of the liquid to strike the wall at the edge of the closing cavity is deflected laterally into the cavity. The effect is cumulative, until the last remaining fraction of the void is filled with greatly augmented velocity by liquid coming from all sides. The wall is deformed locally outward exactly as though struck from the inside and failure may develop a pattern of breakup which is exactly analogous. (But note that because the blow is struck from inside the container, radial cracks lead on its outside surfaces and circumferential members lead on the inside, with cantilever curl toward the outside.) This can occur when flasks of water, sterilized and sealed in autoclaves, are suddenly opened in an inverted position. Air rushing in from the atmosphere propels the liquid upward to close the "bubble."

With pasty or viscous liquids, failure pops out an oval fragment, known in the glass container industry as a "mouse hole" (Fig. 9–8). It is said to be encountered in warehousing some foods which have been packed at steam temperature in glass jars, leaving the head space cavity under near-vacuum when cooled to room temperature. When a case of these jars is tossed on top of another, the jar contents are first impelled to the closure end (top) and then, on rebound, they are moved quickly to the bottom where their momentum is concentrated at the site of the last of the "bubble" there to close.

9.7 Thermal Failures

Thermal failures are the most difficult to discuss in a general way. Among those factors which control the origin and development of cracking are:
1. The temperatures of heat source and heat sink (one of these is the specimen).
2. The area of the specimen participating in heat transfer.
3. The thermal properties of heat source and heat sink, especially their thermal diffusivities. At high temperatures, this can involve radiative heat transfer.
4. Resistance to heat flow across the boundary between heat source and heat sink.
5. The thermal expansion of the specimen, including discontinuous expansion or contraction as a result of phase inversion.
6. The creep properties of the specimen.
7. If the aggressive medium is a gas or liquid, its velocity and angle of incidence. If it is a liquid, its boiling point and heat of vaporization.
8. The presence of flaws as potential crack origins.

Where only thermal stress is involved, tertiary Wallner lines should be absent, except in the proximity of the origin where the "snap" accompanying the start of cracking may generate faint ones. Note, however, that in many cases the chilling agent may be brought

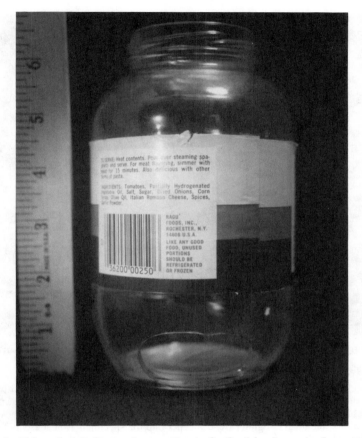

Fig. 9–8. "Mousehole" (bottom) popped out of a food jar by water hammer.

into contact with the specimen violently enough to combine the features of impact with those of thermal failure.

Because of the great variety of thermal failures, they cannot all be covered in the following sections, but a few common examples may be useful to show the principles to be encountered in dealing with a particular case.

9.8   Sudden Chill from a Fluid Medium

A hot specimen subjected to chilling by sudden contact with a cold liquid medium as, for example, when a tumbler hot from the dishwasher is filled with ice water (Fig. 9–9), undergoes prompt shrinkage in its near-surface layer. That layer is therefore placed in a condition of tension parallel to the surface. The interior develops a balancing compressive stress, which is at first negligible in magnitude because it is distributed over a relatively

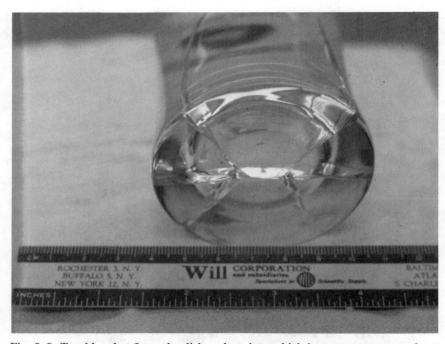

Fig. 9–9. Tumbler, hot from the dishwasher, into which ice water was poured.

great thickness. Failure therefore occurs in the form of a shallow crack which extends rapidly along the surface and perpendicular to it. In doing so, it does not accelerate appreciably, since lengthening of the crack does not increase the stress at the crack tip. In the meantime, the tensile region extends deeper into the specimen as heat escapes from it to the chilled surface, and the crack deepens accordingly.

The history of this crack development can be read on the crack-generated surface. Next to the chilled surface the crack develops only faint tertiary Wallner lines (from the initial snap) and such primary Wallner lines as arise from surface flaws. Precavitation hackle lines and a cavitation scarp show slow acceleration through the depth, followed by a Sierra scarp as the crack arrests and resumes. Several generations of this sequence of cavitation and deceleration may appear. This suggests that sometimes progress through the thickness is partly controlled by access to water and is therefore rhythmic in nature as the crack alternately leaves the water behind and then slows or arrests in the decreasing tensile stress field until water catches up and aids its progress.

If the temperature difference between the specimen and the quenching medium is very great, crizzling may occur in the form of fine checks following a tortuous path and, after extending only a short distance inward, turning to run nearly parallel to the quenched surface, separating flakes.

Liquid droplets on a hot surface may cause simple checks or, depending on their size, may develop surface flaking as noted above.

Chilling of a hot specimen by a less efficient medium, such as a current of air, tends to develop tension more slowly, in a deeper layer, since the boundary resistance to thermal transfer allows time for heat to escape from a greater depth. Failure in this case occurs with greater stored energy, and cracking through the thickness is more rapid, while mist hackle may be seen in the chilled side of the crack surface. Crack initiation may be delayed for an appreciable time after the onset of chilling, but the cracking then occurs with considerable violence because the strain energy at the instant of failure is available from much more of the specimen thickness.

9.9   Sudden Heat

If a specimen surface is suddenly subjected to heating, that surface expands and develops a compressive stress parallel to the surface. The interior is thereby brought under tensile stress. Because flaws are less likely in the interior than at the free surface, and because a buried flaw is only half as effective in initiating cracking as one of equal size exposed at the surface, sudden heat leads to failure less frequently than does sudden chill.

Testing by sudden heat is attractive as a method for evaluating the intrinsic strength of materials, undisturbed by surface peculiarities such as asperites, firing "skin," or machining damage, and unaffected by atmospheric influences. Proof testing by sudden heat is used to reject articles of glassware that contain dangerous "stones."

Sudden heating of a brick brings about the fastest temperature rise at the corners, where access to incoming heat is provided by three surfaces, whereas two surfaces supply heat at an edge and only one at the face. The resulting expansion sets up a steep strain gradient which tends to shear off the corner. There is a corresponding tendency, but a lesser one, to shear off the edges, and last of all to spall off slabs from the faces.

9.10   Delayed Thermal Failure

Failure from sudden heating does not always occur during the heating process, but may result only later, after subsequent cooling and in the absence of thermal gradients. This can take place when the sudden heat is strong enough to raise the surface temperature into the creep range. The high compressive stresses in the surface region are thereby partly relieved by creep; when cooled to uniform temperature, the surface shrinks more than the interior and develops tensile stress equivalent to that portion of the compressive stress which was relieved by creep. In many cases this may be sufficient to cause spontaneous failure.

A steady-state thermal gradient may develop failure by the same mechanism, the crack originating in the initially hottest region only when the system is cooled to ambient temperature.

Note that in both of these cases creep, a nonbrittle phenomenon, is pivotal to the failure. Nevertheless, the failure may take place below the creep range in the regime of brittle fracture, and the crack morphology is that of a brittle material.

9.11   Failure in Coatings

The stress distribution across a coating and substrate depends on whether the interface between them is constrained, as when the substrate is coated equally on both sides or when the specimen is tubular, and on whether or not there is interaction between the two. In this case the stress in the coating is constant across its thickness and the stress in the substrate is also constant but of opposite sign.[4] In such systems cracking is most likely

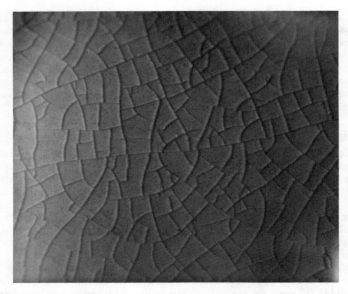

Fig. 9–10. Crazing of a whiteware glaze as the result of moisture expansion of the underlying body.

to occur first at the coating edge, starting into the compression partner at an angle of 45° and curving to run parallel to the interface. The next crack begins in the tension side perpendicular to the interface and then extends into the compression member a short distance before branching to run in it parallel to the interface.

If the interface is unconstrained, bending may occur, bowing the specimen toward the tension member. This relaxes the stresses in each to such an extent that the stress sign may actually reverse across one or both partners. By this mechanism a coating that is expected to be in compression may actually be in tension at its free surface. Cracking may thus be spontaneous in what had been designed to be a compressive coating.

Crazing in glazes on ceramics is a special case of poor fit between a coating and a substrate. Below the strain-point temperature of the glaze, the contraction of the glaze should be lower than that of the body in order to finish with the glaze under relative compression. (But too high a compression must be avoided lest the glaze "shiver," that is, pinch off in spalls at the corners and edges.) Yet even if a desirable balance of stresses is achieved at the end of the firing cycle, crazing may still be occasioned by slow expansion of the body through reaction with moisture. Crazing of the glaze as a result of moisture expansion of the body is illustrated in Fig. 9–10.

Because the crack runs perpendicular to the principal tension at the crack tip, the cracking pattern provides a direct indication of the stress distribution in the specimen at the time of crazing. This effect is especially useful in connection with glazed specimens whose glaze has crazed; the craze pattern yields a complete record of the stresses exerted on it by the body. Craze crack directions are perpendicular to the local principal tension directions and their abundance is an indication of the local stress magnitude.

This suggests the use of high-expansion glazes in process development. Spontaneous crazing when cooling in process will indicate the residual stress systems set up by anisotropic contraction of the body or by uneven cooling. These crazes can then be dyed, preparatory to stimulated-service testing to indicate the service stress distribution by means of the orientation and abundance of new craze cracks. (A similar procedure, in which the specimen is coated with a brittle lacquer, is used with metallic specimens in simulated-service testing, but such lacquers are generally too elastic for use on most brittle materials.)

10   Examples in Practice

Practical examples include the common conditions of failure previously described. They frequently involve combinations of conditions, however, and some of these conditions will be discussed to illustrate the principles to be employed in disentangling the several elements present. Often, too, other principles of science and of common sense come into play if accident reconstruction is involved.

A single sign may possibly be sufficient to establish, in the words of the law, "reasonable scientific certainty." But, wherever possible, the analyst will search for a body of mutually confirming indications until he or she is convinced that absolute scientific certainty has been approached.

It depends on the question. If we are asked whether a certain bottle was broken by internal pressure alone or by an external impact, an explicit answer can be given. If, on the other hand, we are asked how long before the incident the fatal flaw was introduced, we may have no scientific basis to form an opinion (but if other history is provided, limits may be assignable). We can say whether a certain window pane broke from heating of the center, as distinct from having been broken by wind pressure, but often we are unable to tell whether the origin edge flaw was introduced in manufacturing, in transporting, or in glazing.

10.1   The Cleaning Solution

A solution of 70% ether, 30% ethyl alcohol has sometimes been used to clean delicate instruments. Its flammability is not its only peculiarity, as a case in Florida illustrates.

A technician sent his assistant to get a gallon of this solution for use in his home workshop, located in the garage attached to his house. It was dispensed for her into a four-liter bottle from a stock maintained at 68°F. She drove back with it in her car through the 98° heat. The trip lasted about an hour. The workman testified that as she reentered the workshop the bottle burst spontaneously, drenching her clothes with the contents. The fumes drifted along the floor, reaching the pilot light of a clothes dryer; the garage and attached house went up in flames, and the woman died of her burns some six weeks later.

Analysis showed that the bottle had burst from internal pressure, the crack origin being near the center of the base (common for bottles of large diameter). It was marked with prominent subcritical cavitation hackle and a cavitation scarp. Fragments included some with a strong strain pattern, visible under the polariscope, indicating the application of intense heat. Others showed a crizzled pattern from thermal shock and a Sierra scarp typical of quenching with water from high temperature.

The physical properties of the solution can be calculated from those of the constituents. The heat capacity is barely one-half that of the same weight of water, while the weight of one gallon of solution is less than three-fourths that of water; both showed that the rate of increase in temperature in the same warm environment would be very much faster for the solution. At the same time, the thermal expansion of the solution is approximately seven times as great per degree of temperature rise. Thus, the one-and-one-half inches of head space in the bottle, prudent for water, would be far too little to permit harmless expansion of the solution in transit, and pressure increase would be anticipated. Accordingly, the theory that the bottle burst from internal pressure was substantiated. The stored strain was accounted for by the fire that ensued, and the thermal quenching that followed came from the firemen's hoses. Even the presence of the sub-critical cavitation hackle and cavitation scarp at the burst origin were accounted for: The technician testified that his assistant was carrying the bottle with one hand on the neck, with her other (perspiring) hand under the base, supporting the load.

10.2   A Case of "Water" Hammer

The principles of fractographic analysis must be interpreted with common sense. A case in point involved the fracture of two sight glasses in a transfer line delivering pressurized sulfur dioxide from a railway tank car to a plant storage tank. During delivery the sight glasses in the transfer line ruptured. In both cases the origin of cracking was on the outside surface, at the center of the glass disk; a symmetrical array of cracks radiated from there outward in both directions to the edge.

These were typical pressure bursts, not totally unexpected, since the equipment had seen a good deal of service. But why should both glasses fail simultaneously? When the first glass broke why didn't leakage of the contents lessen the pressure on the other? That could be explained only if the pressure rise was so abrupt that leakage had no time to relieve pressure.

The suddenness of pressure rise could be explained by a difference in the valve settings. The entry valve was fully open. A tiny opening in the exit valve would be enough to permit sulfur dioxide gas to pass without appreciable pressure drop. But when a slug of liquid followed, passage was impeded and its kinetic energy was expended against the walls of the transfer pipe—and the innocent sight glasses.

10.3   A Blow from Within

Sterilized water for hospital use is often provided in glass bottles, sealed under vacuum with a rubber cap which can be stripped off quickly for emergency use. Perhaps too quickly! In the operating theater of an eastern hospital a nurse reached for such a bottle only to have it burst, scattering fragments of glass into the patient's incision. Defective glass?

The pattern of cracking in the reassembled bottle was typical of that of a bottle burst by a blunt impact against the outside of the sidewall. But there were differences: The cracks radiating from the center of impact led at the outside surface while transverse cracks led at the inside. Evidently the impact was against the inside surface! How could that have occurred?

The answer lay in the suddenness with which the closure had been removed, together with the attitude of the bottle at that time. With the bottle in the inverted position the water surged to the top end, and on removal of the cap the pressure of the atmosphere drove the water into the vacuum "bubble" where its force was progressively amplified as the bubble closed. It was a classic case of water hammer. No wonder that nurses are warned to open such bottles only when the bottle is exactly erect!

10.4   The Great Australian?

The pop bottle was alleged to have exploded spontaneously, of course. Certainly it had burst, as the pattern of cracking in the reassembled pieces showed (forking angle less than 90°, circumferential crack elements leading at the inside surface). But cracks joined the burst center to the rim at two points—unusual for a spontaneous burst—and moreover the cracks spread from the rim downward and had served as leader cracks, initiating the burst.

It was then alleged that the rim cracks were introduced during the process of applying the aluminum roll-on pilfer-proof closure in the bottling plant. Doubt about that explanation was cast by the presence of clearly outlined indentations in the aluminum cap opposite the points of origin of the rim cracks. (A mirror on a slim rod made it possible to examine the glass within the cap using the stereoscopic microscope.)

It seemed possible that a household tool had been used on the cap, and that it had dented the cap during the attempt to loosen it, breaking the glass rim in the process.

Witnesses testified that no such tool was on the premises. How, then, could the physical observations be reconciled with that testimony? The answer was never provided, but it should be noted that the plaintiff had a powerful jaw and an uncommonly fine set of teeth.

10.5   Hot or Cold?

A type of water faucet handle, common in the beginning of the century, consisted of a porcelain sleeve cemented over a bronze spindle that extended from the valve stem. Fracture of the porcelain led to many injuries, because its sudden failure presents the sharp end of the spindle to the hand or wrist.

Fractographic analysis showed the origin of cracking in several cases to have been at the end of the porcelain sleeve where it contacted the bronze spindle. That is not surprising, since the more compliant bronze could be expected to flex, peaking stresses in the porcelain there.

Examination of many faucets showed that it was always the hot-water faucet handle that had failed or had cracked. The explanation could lie in the thermal expansion mismatch between bronze and porcelain. On reflection, however, another cause might be suggested by anyone responsible for routine maintenance around the home. The washer in the hot-water faucet always wears out before the cold one, and so more and more twisting force must be applied to keep it from dripping. Eventually, a crack forms, and in repeated use it elongates until catastrophic failure occurs.

10.6   Tale of Two Teapots

A shop in an eastern city of the United States received a shipment of cheap teapots and soon sold two of them. On a Sunday morning the first was used by a father, assisted by his four-year-old daughter, in preparing breakfast in bed for his wife. The bottom fell out of the teapot, drenching the little girl with scalding tea and seriously scarring her.

The next day a mother was making an afternoon pot of tea when her little daughter was scalded in just the same way when the second of the teapots failed.

A neighborhood lawyer had the teapot fragments examined, and he sued the storekeeper on the strength of an opinion that the teapots were defective (the ornamental flower pattern was somewhat smeared!). The case was settled out-of-court for a healthy sum. The storekeeper then sued the Canadian manufacturer to recover.

The fractographic markings on the crack surfaces of a porous earthenware are notoriously difficult to detect, but the markings on the glaze of the teapot surfaces were more

Fig. 10–1. Crack origin lies in inner surface of the glaze (center top); no fracture markings are visible in the underlying body. ×70.

tractable (Fig. 10–1). The glaze on the inside had its fracture origin opposite the foot in the base; the glaze on the outside showed that crack spread had run outward from the body. Assuming that the failures were thermal in nature, this indicated that the hottest region was in the foot, the inside of the pots being relatively cool.

Evidently the teapots had been put on the stove to boil water and had then cracked, allowing the bottoms to fall out when they were removed to pour out the tea. Inasmuch as this use of a teapot to boil water constitutes misuse, the verdict was returned for the manufacturer.

10.7   A Dish of Lasagna

The lady complained that she had baked a dish of lasagna for dinner, had stored the remainder in the refrigerator, and when she lifted the dish out for lunch next day it broke for no reason at all. At first glance it appeared that the crack origin was probably somewhere along the crack bisecting the baking dish diagonally (Fig. 10–2). It wasn't; it was on the upper surface of the short crack which intersected it near the corner. The origin was part of a short, strongly curved segment. Could it be a thermal check received during manufacture?

(A)

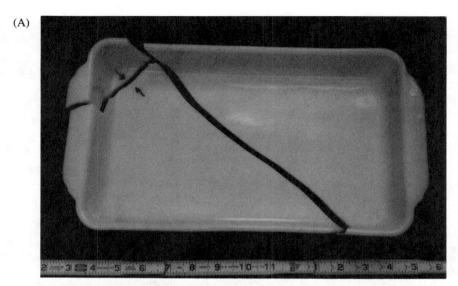

(B)

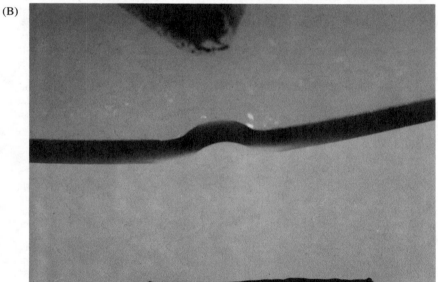

Fig. 10–2. Baking dish fracture showing (A) cracks, with origin marked by arrows; (B) crack profile at origin showing a ring-crack segment and beginning of subsurface Hertzian cone development; ×100.

The profile had not only curvature, characteristic of chill checks; it was exactly a portion of the arc of a circle. Significantly the circle was continued into the surrounding glass at both ends of the arc (not visible in Fig. 11–2). The failure had begun in a ring crack—the first stage of a Hertzian percussion cone.

The impact which had led to failure could have occurred at any time after the baking dish was formed. But, as the dish had been in service for some time (numerous scratches and stains testified to that) it was reasoned that it was probably abuse by the plaintiff that was to blame.

10.8   The Fatal IV Bottle

A hospitalized young woman was given an intravenous feeding solution; she soon developed a fever and died. Analysis showed that the IV bottle's glucosaline solution contents had been contaminated with the pathogenic bacteria *aerobacter cloacae*.

A long crack in the heel of the bottle suggested how the contamination had entered. It was not so clear how it had been overlooked by the nurse on duty, who had been instructed to inspect IV bottles carefully before administration. Had she neglected to see that the rubber seal was not bowed inwards by the vacuum within? And had she failed to listen for the hiss when the closure was penetrated by the delivery needle? Yet she testified that she had not been negligent in any of those respects.

It is one thing to detect a crack in a dry, empty bottle and quite another matter to see a crack in a water-filled one. Viewed in reflected light, the dried crack showed seventh-order green interference colors, indicating a crack width of 1.9 μm. In such a crack it can be shown that air from the atmosphere cannot leak in against the capillary force of a dextrosaline solution and so would not be admitted to break the vacuum. Contaminants in aqueous suspension are not opposed by capillarity and can be readily admitted.

Evidently the cracked bottle had come into contact with dirty water which had intruded through the crack and had contaminated the contents without disturbing the vacuum. The nurse was therefore held blameless, but the pharmaceutical firm, whose inspection procedures were questionable, was held to be responsible.

10.9   More than Skin Deep

A manufacturer of heavily grogged earthenware sewer pipe was worried about their marginal strength and so gave his R&D people the task of strengthening them by adjusting the level of compressive stress in the glaze. Many experiments later they reported that they were unable to achieve any improvement, even though they raised the glaze compression to values close to the limit where shivering began.

This finding was so disheartening to the manufacturer that he sought outside help. As in any strength problem, fractographic determination of the origin of failure was indicated. The origin of cracking under the standard loading was located, as expected, 45° around the circumference from the loading point.

But closer examination showed that the origins lay not in the glaze, the region of highest test stress, but at the edge of grog particles beneath it in the body. Obviously, then, it was pointless to experiment with the glaze composition. The body was the weak link in the system, and work was redirected to the composition and particle size of the grog.

10.10   A Faulty Tool

A plant was engaged in the manufacture of unglazed fuse bushings, designed to contain the explosive power released upon failure of a fuse wire under currents of hundreds of thousands of amperes. They were only marginally acceptable, so efforts were made to raise their strength. Within two years the R&D laboratory had developed a new ceramic composition which, though containing more costly ingredients, showed a major improvement when subjected to the standard internal-pressure test.

Should the new composition be introduced into manufacture? An outside opinion was solicited, and fractographic analysis of test failure of the normal bushings and the newly developed bushings was performed.

The new high-strength bushings were observed to fail in test from minor irregularities in the interior (bore) surface, where calculations showed that the maximum tensile stress was indeed developed. Failure in production bushings, by contrast, invariably began in the mounting screw holes of the end surfaces. Dull or burred tools had left checks in the production specimens, a fault that was avoided in the more carefully machined laboratory products.

The expense of the composition change was not justified. Greater care in the machining of production parts accomplished the desired improvement in test performance.

10.11   Something New under the Sun

Solar energy is an attractive prospect to a people anxious to continue their enjoyment of electric blankets and air conditioners. One concept involves reflecting the sun's rays in an array of mirrors so that they may be brought to bear on a target where the heat is absorbed and transferred to a steam turbine. But the same sunny areas of the country where solar energy is feasible are also subject to hailstorms which might threaten the necessary mirrors.

Can a plate glass mirror withstand bombardment by hailstones? Simulated tests with spheres of ice shot randomly against such mirrors, at velocities approximating the terminal velocity of hailstones, gave disappointing results. The mirrors were shattered at many points, with sunburst patterns of cracks characteristic of localized impact.

Fractographic analysis gave some room for optimism. Each impact failure had been triggered by a leader crack coming in from a flaw at the edge.

Thermal effects were also of concern, for the same reason that large plate-glass windows are subject to occasional failures from heating by sunlight. Indeed test arrays of solar energy mirrors, carefully protected from storms, showed a high incidence of cracking.

Again every failure was determined to have originated at a defect in the edge!

The lesson was clear. Inherently not endangered by hailstorms or by the sun's heat, solar collectors must be of glass that is properly cut and protected from the introduction of edge defects from careless handling.

10.12   You May Open It

What started out as a nice gesture at a dinner party became a case for the paramedics and a lawyer. The waiter started to draw the cork from a wine bottle, stopped, got a fresh start, and tore the neck from the bottle, cutting his hand severely.

He was using a common type of levered cork puller, in which the corkscrew is first inserted; a metal shoe rests on the bottle rim while the corkscrew is pulled upward by a lever. The whole apparatus folds together neatly, to be pocketed for convenience.

In his first attempt he evidently heard a cracking sound which led him to rotate the bottle a quarter turn before trying again. The crack which he had started during his first try then ran downward an inch. It forked, and ringed around the bottle neck in both directions, exposing a jagged lip which cut the tendons in his wrist.

It is at least a sobering warning against applying load to parts which are known to be cracked, and it shows again how fractographic analysis can be used to reconstruct a traumatic event in which the witnesses have at best a confused notion as to the details of what has occurred.

10.13   An Ingenious Test

Tensile strength at high temperature was being measured by an ingenious method. The specimen was prepared as a bushing down whose bore a graphite heater rod was led. At the test temperature a measured current of electricity was run through the heater rod, expanding the inside radius and applying hoop tension to the outside. The temperature difference between core and exterior could be calculated from the energy pulse in the graphite rod. Failure was indicated by interruption of continuity in a metal strip, plated on the outside. Successively higher currents were imposed until failure was observed to have occurred and the tensile stress could then be calculated.

If the calculated tensile strength values had shown a reasonable trend with the test temperature, all would have seemed well. They did not.

Fractographic analysis showed that some of the specimens had failed from an origin at the outside, as expected. Some, however, showed the origin of cracking to have been on the inside surface, a totally unexpected result. Evidently those specimens had not failed under the test-imposed thermal gradient, but on cooling. At the high temperature, the compressive stress at the core had been partly relieved by creep and, when cooled, this originally compressive region came under tensile stress high enough to cause failure.

A further observation served to distinguish the two cases. Cracking from an origin on the outside relieved the imposed stress, and the cracking event was complete when failure was noted. On the other hand, cracking from the core outward resulted in additional tension in the opposite side of the core and a second crack resulted, completely splitting the bushing. Since this second crack included a bending component, it ran straight along a radius at first, but then developed cantilever curl.

Worst of all, the stress relief in the core by creep in one cycle introduced a state of residual stress which made it impossible to calculate the tensile stress at failure in a subsequent cycle. The test was abandoned.

10.14   A Pressure Vessel Model

It was scarcely practical to test to failure a hundred-foot concrete containment vessel with walls eight feet thick to see if the design was correct. It was thought preferable to test a plastic model made carefully to scale. The model failed at a disappointingly low pressure.

The cause of failure was evident from the location of the crack origin. It was found to lie in the tubulation provided for application of the test pressure, a fitting which simply screwed into a drilled-and-tapped opening. Since this was the only detail in which the model did not correspond to the actual containment vessel, there was no cause for alarm. A new model was constructed with more attention to the conformation of the pressure inlet. The new test established that the containment vessel concept was sound, and it was constructed with confidence.

10.15   The Cold Truth

She said it just "blew up in her hand," cutting her tendons at the wrist. It was a bottle of citrus fruit concentrate which she alleged must have spoiled and developed pressure within, even though she had stored it carefully in the refrigerator.

Certainly it was a classic burst pattern—origin in the sidewall, with cracks fanning vertically upward and downward and a maximum forking angle of 90°. The origin was on the outside, developing a dead-flat mirror without cantilever curl or inside spalling.

But there was something very wrong with certain of the observations. The fragments had not been blown out but were retained within the fragile foamed-plastic jacket. And there was no indication of spoilage of the contents. Where had the internal pressure come from?

She had stored it in the refrigerator all right, possibly in the freezer compartment. There it had evidently frozen, with consequent expansion of the fruit juice and pressure development against the walls of the container.

A dozen fresh bottles were burst with gas pressure and a dozen were stored in the deep freeze. Gas pressure resulted in flying fragments and shredding of the foam plastic jacket in every instance, while the frozen bottles invariably showed a typical burst pattern with the jacket intact.

10.16   An Ancient Art Explains a Modern Catastrophe

Glued-glass is a time-honored technique for generating brilliant, fanciful patterns on flat glass for window panes and the edges of mirrors. The process is simple: The glass, slightly roughened, is coated with glue or gelatine which is allowed to set and then is baked to dry it thoroughly. The drying is accompanied by shrinkage which develops a tensile stress in the coating and a compressive stress in the base glass. Shallow chips are thereby torn from the glass in a random pattern resembling frost "flowers."

The strength of the pane is scarcely reduced by the process when properly carried out, since each chip-out is completed smoothly and provides no incipient fracture origins.

The process was called to mind in explaining the mysterious failure of the glass spandrels in a large building in the Midwest. The glass panes were found to have been accidentally spattered with weld metal. Weld spatter on the windows of a high-rise building is, admittedly, an objectional cosmetic defect. But on the tempered glass spandrels between the vision panes it is tempting to think that it might safely be overlooked as insignificant. In fact it is a dangerous condition that can lead to delayed failures days or even years after installation.

The mechanism of such delayed failure is interesting. When a droplet of hot weld metal strikes the glass its action is twofold. To begin with, by heating the tempered glass it may bring the temperature above the strain point, so that the tempering stress is relieved locally and the balance of stresses is upset. The protective compression layer is thereby interrupted, and the strength of the sheet is correspondingly reduced. It is not quite so obvious that the drop of metal, now in intimate contact with the less thermally expansive glass, can develop tensile stresses bordering the drop, in the initially compressive region of the glass. As a result, cracks start inward at approximately 45° to the surface, turn, run parallel to the surface, and meet opposite the center of the drop. In this way they reduce the effective thickness of the glass and so lower strength somewhat. But where chipping has begun but is not completed, in each loading of the structure thermally or by high winds, the cracks may change direction and extend a little farther, until eventually they reach the tensile zone and the glass fails spontaneously and catastrophically.

10.17   En Garde!

A nationally ranked lady fencer sued a manufacturer, claiming that his glass meat thermometer had broken when inserted into a rolled roast of beef, severing her wrist tendon and ending her fencing career.

The thermometer had failed at a notch used to secure the temperature scale, and a long razor-sharp sliver of glass had been released parallel to the stem.

Reenactment of the scenario on a testing machine confirmed that this was the expected mode of failure, but the force necessary was surprisingly high.

The lady testified that she had been expecting guests for a 6:30 dinner. Arriving home from her work at approximately 4:30, she had taken the dog for a walk. When she returned at a little after 6:00 she went down to the oven and attempted to insert the thermometer. It broke, causing her injury.

A woman thought of the right questions for the defense attorney. Did she always walk the dog for an hour and a half? Was everything ready for her guests: Was the table set, the salad tossed? Was she dressed for her dinner party? Had anything happened on her walk to upset her? No to all of these except the last. Her disturbed state of mind and her haste, combined with her exceptional ability as a fencer, resulted in a vicious jab at the roast, loading the thermometer to an extent that was judged to be unforeseeable by the manufacturer.

10.18   A Gradient in Temperature

Tempered-glass cylinders were failing to meet impact-strength tests. The reason was clear when the fragmentation pattern was examined. Near one end, the dicing pattern characteristic of well-tempered glass was in sharp contrast to the other end, which broke into the large pieces characteristic of insufficiently tempered glass (Fig. 10–3).

Improper preheating of the cylinder prior to the chill was inferred to be responsible. Because of a faulty burner, the preheat furnace proved to have a steep temperature gradient from top to bottom, and correction of this condition cured the problem.

10.19   The Picture of Dr. X

It was just a framed picture on his wall—next to one of his wife and children. It showed a specimen—a bushing—split in two. Its story? It was a uranium dioxide fuel element, tested to determine if it could be a useful fuel element for a nuclear gas-cooled power reactor. It had been sheathed in a water-cooled aluminum shell and had been inserted in a nuclear pile to raise its temperature by the fission reaction and so to determine whether it could survive the thermal load it would receive as a nuclear fuel. It had failed.

But it was an intriguing picture. One side showed a simple planar crack surface. The other was initially planar near its core but it was curved out of plane toward the outer surface. Had it really failed during the power test?

Under power, the core of the bushing would be hottest, would expand, and would

Fig. 10–3. Glass cylinder tempered to dicing stress at the lower end, insufficiently tempered at the upper end.

develop a compressive stress. The exterior, in contact with the water-cooled cladding, would remain cool and would develop a tensile strength to balance the compressive stress at the interior. The bushing should therefore have cracked, if at all, from the outside in. Such a crack would have relieved the tension, and the failure event would be over.

But evidentally it didn't fail under power. The compressive stress at the interior would cause stress relief through creep. On cooling, the initially hot compressive region would have furthest to cool, would shrink relative to the outside, and would pull itself into tension. Cracking from inside out could be expected to increase the tension at other points along the inside—especially opposite the initial crack—and a second crack might extend from the inside, developing cantilever curl before exiting.

So perhaps the test assembly had merely been withdrawn too quickly from the test pile! Retrieval of the specimen from storage for examination confirmed the location of the

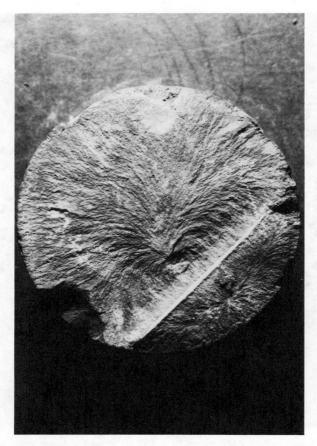

Fig. 10–4. Broken end of a drill core with two origins of cracking within the broken surface.

crack origin at the bore. The test had actually demonstrated the soundness of the fuel element concept.

10.20   The Tell-Tale Origins

Drill cores, drawn from potentially oil-bearing rocks, were being recovered in short, broken lengths, leading to the inference that the deep strata were extensively shattered along a plane perpendicular to the drill hole. Fractographic examination led to a serious modification of that view. It became clear that the origins of cracking in almost every case lay within the core (Fig. 10–4).

It was recognized that the breaks in the core were effected during drilling, or during withdrawal of the drill string, and that the rock strata themselves were unbroken.

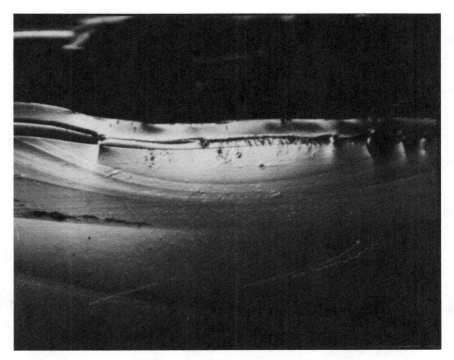

Fig. 10–5. Burst origin in the outside base of a large glass container is a reentrant at the edge of a lapped baffle mark. × 100.

10.21 The Bursting Bottles

The bottles were bursting spontaneously. The origin was always in the outside surface of the base, not far from the heel. What was responsible for the flaw there?

The origin flaw looked simple enough; it was a shallow fissure such as that from a chill check (Fig. 10–5). Inspection of the free surface was more revealing; it showed that the fissure was a long detail, curved parallel to the heel of the container, and technically known as a lapped baffle mark. A loose fit between the preform mold and its baffle plate had permitted an extrusion of glass between them; in the blow mold this was folded flat, forming a sharp, dangerous reentrant angle with the ware surface. Remachining of the preform mold parts to give a closer fit eliminated the difficulty.

10.22 Panic in the Gym

The immaculate basketball floor was suddenly strewn with fragments of glass raining down from one of the recently installed high-powered lamps in the ceiling. Closer examination showed that several of the cover glasses were cracked and evidently prepared to contribute further debris.

A glance at the cracked covers revealed the cause of the trouble. The cracking pattern

Fig. 10–6. Tempered cover glass of a high-powered lamp which cracked spontaneously as a result of intense heating in service, unbalancing the protective tempering stresses. Origin is at the center of the "butterfly" at the center.

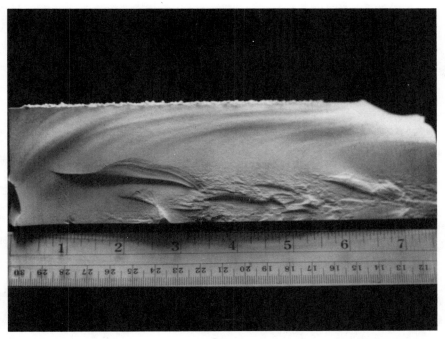

Fig. 10–7. Broken section of a large porcelain tube which cracked spontaneously during machining of the end surface. Mist hackle on the inside surface (below) showed that a high stored tensile stress was present there as a result of too-fast cooling through the annealing range of the glass phase.

near the edges was of the intricate "dicing" type typical of tempered glass, but near the center it consisted of large fragments characteristic of annealed glass. The center of the tempered glass had been earlier "stress-relieved" by the heat of the bulbs, so that their enhanced strength had been lost (Fig. 10–6) and they cracked when next turned on.

Chemical analysis showed that the ratio of MgO to CaO had resulted in an abnormally low strain point for this glass, so that the cooling fans in the lamps could not prevent heating to a temperature at which softening of the glass allowed relief of its protective stresses.

10.23   The Tube

A porcelain manufacturer took a special order to provide a tube of unusually large dimensions, requiring a lengthy series of operations. The drying and firing times alone took eight months. The last step consisted of machining the end surfaces flat and smooth. The tube cracked from end to end! With no explanation for the failure and faced with the prospect of repeating the entire process with the possibility of another broken product, the manufacturer asked if fractography might provide some reason for the initial failure.

Examination of the crack-generated surface showed that the crack had originated at the partly machined end and had run at terminal velocity, leading strongly at the inside (Fig. 10–7).

It could be inferred that the porcelain, under compressive stress at the outside and tensile strength at the inside, had been cooled too rapidly through the temperature range where the glassy phase was passing its yield point. The cooler outside wall had become stiff, while the warmer inside could still relieve the differential-shrinkage stresses by creeping. When the tube was approaching room temperature, the hotter inside had further to cool and to shrink, pulling itself into a dangerous tension.

11 The Expert Witness

The fractographer may be called as an expert witness in a case at law. There is a marked difference between what the layman thinks an expert is and what is considered an expert in a court of law. In the law an expert is simply someone who knows more about a certain subject than the average layman.

To be qualified as an expert witness is to have special privileges. The lay witness is restricted to testifying only about matters of fact—what he has seen, what he has heard (but not, of course, merely what he has been told), what he has experienced firsthand. The expert witness, on the other hand, is permitted to express opinions.

That is not to say that the jury must believe him. Therefore, he is wise to present the jury with a carefully reasoned explanation of the grounds for any opinions expressed. For this purpose, sketches, photographs, and models are usually admissible and in many courts television presentations are now permitted. Sketches, made in the courtroom and carefully labeled, are particularly effective.

Testimony is not rehearsed, of course. Nevertheless, it is helpful for the expert witness to go over with his lawyer what questions he plans and what sort of answers will be given. To follow and direct his case intelligently, the lawyer needs to understand the technical points involved and he needs his expert witness to coach him toward that end.

At the very least, it is a chance to agree on the vocabulary. For example, a witness who specializes in fracture mechanics will be apt to explain that a certain fracture began at a flaw. If he is for the defendant, his lawyer would prefer him to say that fracture began at a singularity, because the word "flaw," to the layman, is dangerously close to the word "defect," and so it sounds like a tacit admission that the product was unsafe for its intended purpose.

The lawyer does not want to ask in court any question to which he doesn't know the expert's answer, so he will ask him all possible questions before going into court.

The opposing lawyer is also permitted to ask questions before the actual trial, in a proceeding called a "deposition in discovery," to which he is entitled under the general understanding that both sides should have access to all the facts, as known to witnesses, and to all facts to be relied on in opinions to be expressed by expert witnesses.

Something should be said about the deposition. The expert witness's own lawyer will be present as well as his opponent (who will do most of the questioning) and a legal secretary. You are warned that you are testifying under penalty of perjury. The record of your testimony in deposition is taken down verbatim and may be quoted at trial and used as the basis for cross-examination. So it is prudent to weigh your words as carefully as if you were in court.

Now you are in court. You have sworn to tell the truth, the whole truth, and nothing but the truth, and your own lawyer begins the questions. Do you recognize the physical evidence? By what means to you recognize it? Did you have a chance to study it previously? Did you study it? By what means? Were you able to draw any conclusions? On what did you base those conclusions? What were your conclusions?

Many things are explainable by common analogy, as when we describe extrusion by saying it is like squeezing luncheon meat out of a meat grinder. Of course we don't want to offend the jury by "talking down" to them, but we must explain things as nearly as possible in their language. If we do a really good job at this, maybe the jury will reach the same conclusion that we have reached even before we've told them what that conclusion is.

The PSI system of units has not yet filtered down to the layman. So don't neglect to have any figures you are going to cite convertible into everyday units. For example, if you give strength in meganewtons per meter squared, be prepared to answer how many pounds per square inch that would be.

Up to this point, when your attorney concludes his questioning, or "direct" examination, much of the problem is solved by him. He knows the rules of evidence, does not try to lead his witness, and listens to the answers so that he can follow them up with his own useful questions. He remembers the points that are to be brought out and provides questions that allow them to be made.

Then the opposing attorney gets his chance to ask questions—to cross-examine. It is helpful to think of him as a colleague whose questions will help you to throw light on the case. That is by no means what he has in mind, however. He is likely to try to embarrass and confuse you.

You will get a lot of good advice when you appear: (1) Never volunteer information; answer only what is asked. (2) Be sure that the question is complete before you answer. (3) Make sure that you understand the question. (If you don't understand the question, say so and ask that it be rephrased, not just repeated.) (4) Don't be trapped by "no win" questions like the famous: "Have you stopped beating your wife? Answer yes or no." The lawyer is entitled to demand a yes-or-no answer but you are entitled to explain such an answer.

A common trap comes in the form: "Are you telling the jury, Doctor, that the plaintiff is lying?" Now of course that is not what you are saying. (The plaintiff may have a bad memory, or may have been helped through his story so often that he believes it to be true.) So a simple restatement of your opinion in the case is the safest answer.

During cross-examination the opposing lawyer will try to limit you to answers that give a misleading impression. Your own lawyer can help you to correct these impressions through questions he can put to you in what is called "redirect" examination. It will help him if you can remind him of what points he ought to cover.

For a scientist or engineer, the most difficult habit to break is the tendency to answer only in terms of absolute certainty. In the law, answers are to be given according to the best of your knowledge and belief, in accordance with established principles of engineering and science. In other words, no one is interested in the one-in-a-million possible exception to a statement that is, in general, true.

You may take any materials to the witness stand that you may want to refresh your memory. But if you do take something for that purpose it is well to remember that the opposing attorney may request that it be admitted into evidence and that he may ask you detailed questions about it. It is well to make sure that it includes nothing that might embarrass you.

This question-and-answer exchange is not a dialogue between you and the lawyers. It is actually a conversation with the jury; the lawyer is simply asking aloud the questions that the jury would like to ask if they were permitted to. (They are not permitted to ask any questions at all.) Therefore your answers should be directed toward them.

Juries are by no means the pushovers for the plaintiff's sob stories that is popularly supposed. There have been cases where the jury was literally in tears over the unquestioned tragedy of the plaintiff's suffering, only to give the verdict to the defendant. Conversely, although a certain jury had to be impressed with the thoroughness of the defendant's quality assurance program, they returned a plaintiff's verdict.

The whole thing is really a matter of putting your expertise at the disposal of the jury—with the help of your attorney, in spite of the opposition lawyer, and with the judge to see that there is fair play. The expert witness should avoid making any judgment as to the rights of the case or letting sympathy for his client cloud his mind. It is not for him to decide the issue; it is for the jury alone to determine the facts and who is to prevail.

References

1. The Initiation and Development of Brittle Failure

[1]A. A. Griffith, "The Phenomena of Rupture and Flow in Solids," *Philos. Trans. R. Soc. (London), Sect. A,* **221,** 163 (1920).

[2]C. E. Inglis, "Stresses In a Plate Due to the Presence of Cracks and Sharp Corners," *Trans. Inst. Naval Archit.,* **55,** 219 (1913).

2. The Fundamental Markings on Crack Surfaces

[1]V. D. Frechette, "Markings on Crack Surfaces of Brittle Materials: A Suggested Unified Nomenclature," ASTM Spec. Tech. Publ. 827, Philadelphia, 1984.

[2]H. Wallner, "Structure of Lines on Fracture Surfaces," *Z. Phys.,* **114,** 368 (1939).

[3]William E. Snowden, "Crack Growth in Glass Subjected to Controlled Impacts"; Ph.D. Thesis, University of California, Berkeley, 1976.

[4]W. J. Galloway, "An Experimental Study of Acoustically Induced Cavitation," *J. Acoust. Soc. Am.,* **26,** 849–57 (1954).

[5]T. A. Michalske and V. D. Frechette, "Dynamic Effects on Crack Growth Leading to Catastrophic Failure in Glass," *J. Am. Ceram. Soc.,* **63** [11–12] 603 (1980).

[6]T. A. Michalske, M. Singh, and V. D. Frechette, "Experimental Observation of Crack Velocity and Crack Front Shape Effects in Double-Torsion Fracture Mechanics Tests"; pp. 3–12 in ASTM Spec. Tech. Publ. 745, Fracture Mechanics for Ceramics, Rocks and Concrete. Edited by S. W. Freiman and E. R. Fuller. Philadelphia, Pa., 1980.

[7]S. M. Wiederhorn, "Fracture Energy of Soda-Lime Glass," *Natl. Bur. Stds. Rept.,* **8618,** 1965.

[8]J. H. Varner and V. D. Frechette, "Fracture Marks Associated with Transition-Region Behavior of Slow Cracks in Glass," *J. Appl. Phys.,* **42,** 1983 (1971). C. L. Quackenbush and V. D. Frechette, "Crack-Front Curvature and Glass Slow Fracture," *J. Am. Ceram. Soc.,* **61,** 402 (1978).

[9]J. A. Kies, A. M. Sullivan, and G. R. Irwin, "Interpretation of Fracture Markings," *J. Appl. Phys.,* **21,** 716 (1950).

[10]A. Tsirk, "Formation and Utility of a Class of 'Anomalous' Wallner Lines on Obsidian"; pp. 57–59 in Fractography of Glass and Ceramics. Edited by V. D. Frechette and J. R. Varner. The American Ceramic Society, Westerville, Ohio, 1988.

3. The Pattern of Forking

[1]F. W. Preston, "The Angle of Forking of Glass Cracks as an Indicator of the Stress System," *J. Am. Ceram. Soc.,* **18** [6] 175 (1935).

[2]T. A. Michalske; personal discussion.

[3]F. Kerkhof; p. 203 in Bruch Vorgaenge in Glaesern. Verlag der Deutschen Glastechnischen Gesellschaft, Frankfort/Main, Federal Republic of Germany, 1970.

4. The Seeds of Failure

[1]B. R. Lawn and T. R. Wilshaw, "Indentation Fractures: Principles and Applications," *J. Mater. Sci.,* **10** [6] 1049–81 (1975).

[2]J. R. Varner and H. J. Oel, "Surface Defects: Their Origin, Characterization and Effects on Strength," *J. Non-Cryst. Solids, 19,* 321–33 (1975).

[3]A. G. Evans, "Structural Reliability: a Processing-Dependent Phenomenon," *J. Am. Ceram. Soc.,* **65** [3] 127–37 (1982).

[4]W. E. Milani, J. R. Varner, and J. S. Reed, "Strength-Limiting Defects in Alumina Substrates"; pp. 1697–1706 in High Tech Ceramics. Edited by P. Vincenzini. Elsevier, Amsterdam, 1987.

[5]R. W. Rice, "Pores as Fracture Origins in Ceramics," *J. Mater. Sci., 19* [3] 895–914 (1984).

[6]F. F. Lange, "Structural Ceramics: A Question of Fabrication Reliability," *J. Mater. Energy Syst.,* **6** [2] 107 (1984).

[7]F. F. Lange, H. Shubert, H. Claussen, and M. Ruhle, "Effects of Attrition Milling and Post-Sintering Heat Treatment on Fabrication, Microstructure and Properties of Transformation Toughened ZrO_2," *J. Mater. Sci.,* **21** [3] 768–74 (1986).

[8]F. F. Lange, "Processing-Related Fracture Origins: I. Observations in Sintered and Isostatically Hot-Pressed Al_2O_3/ZrO_2 Composites," *J. Am. Ceram. Soc.,* **66** [6] 396–98 (1983).

[9]F. F. Lange and M. Metcalf, "Processing-Related Fracture Origins: II. Agglomerate Motion and Cracklike Internal Surfaces Caused by Differential Sintering," *J. Am. Ceram. Soc.,* **66** [6] 394–406 (1983).

[10]F. F. Lange, B. I. Davis, and I. A. Aksay, "Processing-Related Fracture Origins: III. Differential Sintering of ZrO_2 Agglomerates in Al_2O_3/ZrO_2 Composite," *J. Am. Ceram. Soc.,* **66** [6] 407–408 (1983).

5. Estimation of Stress at Failure

[1] E. B. Shand, "Strength of Glass—The Griffith Method Revised," *J. Am. Ceram. Soc.,* **48** [1] 43 (1965).

[2]T. A. Michalske; personal communication.

[3]E. B. Shand, "Correlation of Strength of Glass with Fracture Flaws of Measured Size," *J. Am. Ceram. Soc.,* **44,** 451 (1961).

[4]T. A. Michalske; personal communication, 1983.

[5]Victoria E. Henkes, "Microbranching Frequency as A Criterion for Mirror Width"; B. S. Thesis, New York State College of Ceramics, Alfred, 1978.

[6]J. J. Mecholsky, A. C. Gonzalez, and S. W. Freiman, "Fractographic Analysis of Delayed Failure in Soda-Lime Glass," *J. Am. Ceram. Soc.,* **62** [11–12] 577–80 (1979).

[7]J. J. Mecholsky, R. W. Rice, and S. W. Freiman, "Prediction of Fracture Energy and Flaw Size in Glasses from Measurements of Mirror Size," *J. Am. Ceram. Soc.,* **57** [10] 440–43 (1974).

[8]S. W. Freiman, "Brittle Fracture Behavior of Ceramics," *Am. Ceram. Soc. Bull.,* **67** [2] 392–402 (1988).

[9]J. J. Mecholsky, "Fracture Analysis of Glass Surfaces"; pp. 568–90 in Strength of Inorganic Glass. Edited by C. R. Kurkjian. Plenum, New York, 1985.

[10]V. D. Frechette and T. A. Michalske, "Fragmentation in Bursting Glass Containers," *Am. Ceram. Soc. Bull.,* **57** [4] 427–29 (1978).

[11]V. D. Frechette and S. L. Yates, "Fragmentation of Bottles by Impact," *J. Am. Ceram. Soc.,* **72** [6] 1060 (1989).

6. Effects at Inclusions

[1]R.W. Rice and R. C. Pohanka, "Grain-Size Dependence of Spontaneous Cracking in Ceramics," *J. Am. Ceram. Soc.,* **62** [11–12] 559–63 (1978).

[2]F. F. Lange, "Criteria for Crack Extension and Arrest in Residual Localized Stress Fields Associated with Second Phase Particles"; pp. 599–609 in Fracture Mechanics of Ceramics, Vol. 2. Edited by R. C. Bradt, D. P. H. Hasselman, and F. F. Lange. Plenum, New York, 1974.

[3]R. W. Rice and D. Lewis, III, "Limitations and Challenges in Applying Fracture Mechanics

to Ceramics''; pp. 659–76 in Fracture Mechanics of Ceramics, Vol. 5. Edited by R. C. Bradt, A. G. Evans, D. P. H. Hasselman, and F. F. Lange. Plenum, New York, 1983.

[4] H. J. Oel and V. D. Frechette, ''Stress Distribution in Multiphase Systems: II, Composite Disks with Cylindrical Interfaces,'' *J. Am. Ceram. Soc.*, **69** [4] 342–46 (1986).

7. Anisotropic Materials

[1] J. L. McCall, Fracture Analysis by Scanning Electron Microscopy. Metals and Ceramics Information Center, Columbus, Ohio, 1972.

[2] G. A. Wolff and J. D. Broder, ''Cleavage and the Identification of Minerals,'' *Am. Mineral.*, **45**, 1230 (1960).

[3] R. W. Rice, ''Fracture Topography of Ceramics''; pp. 439–72 in Surfaces and Interfaces of Glass and Ceramics. Edited by V. D. Frechette, W. C. LaCourse, and V. L. Burdick. Plenum, New York, 1974.

[4] R. W. Rice, ''Ceramic Fracture Features, Observations, Mechanisms, and Uses''; pp. 5–103 in Fractography of Ceramic and Metal Failures, ASTM STP 827. American Society for Testing and Materials, Philadelphia, 1982.

[5] A. B. Feinberg, ''Dunting in Whitewares''; B. S. Thesis, Alfred University, Alfred, N.Y. 1982.

8. Procedures and Techniques

[1] H.-R. Wenk, editor, Electron Microscopy in Mineralogy. Springer-Verlag, New York, 1976.

9. Common Conditions of Failure

[1] L. Orr, ''Practical Analysis of Fractures in Glass Windows,'' *ASTM Mater. Res. Stds.*, **12** [1] 21–23; 47 (1972).

[2] R. E. Mould, ''The Behavior of Glass Bottles Under Impact,'' *J. Am. Ceram. Soc.*, **35** [9] 230–35 (1952).

[3] H. M. Dimmick, ''Observed Direction of Leader Cracks from Hinge Failures of Glass Bottles,'' *J. Am. Ceram. Soc.*, **35** [9] 235–36 (1952).

[4] H. J. Oel and V. D. Frechette, ''Stress Distribution in Multiphase Systems: I. Composites with Planar Interfaces,'' *J. Am. Ceram. Soc.*, **50** [10] 542–49 (1967).

Index